Of Time and Turtles

Other Books by Sy Montgomery

Of
Time and
Turtles

Mending the World,
Shell by Shattered Shell

SY MONTGOMERY

Illustrations by Matt Patterson

MARINER BOOKS
New York   Boston

OF TIME AND TURTLES. Copyright © 2023 by Sy Montgomery.
All rights reserved. Printed in the United States of America. No part of
this book may be used or reproduced in any manner whatsoever without
written permission except in the case of brief quotations embodied in
critical articles and reviews. For information, address HarperCollins
Publishers, 195 Broadway, New York, NY 10007.

HarperCollins books may be purchased for educational, business,
or sales promotional use. For information, please email the
Special Markets Department at SPsales@harpercollins.com.

FIRST EDITION

Designed by Renata DiBiase

Library of Congress Cataloging-in-Publication Data has been applied for.

ISBN 978-0-358-45818-0

23 24 25 26 27 LBC 5 4 3 2 1

To Dr. A. B. Millmoss
with all my love for eternity

Nature does not hurry,
yet everything is accomplished.

—LAOZI

Contents

Author's Note

Some names and locations have been changed to protect turtle nesting grounds.

Of Time and Turtles

I.

Shell Shock

Pizza Man, a red-footed tortoise

Amid all the other homes on the suburban street—white, beige, gray, pale blue, light yellow—this two-story saltbox stands out. It's a blazing neon green, its flamboyance accentuated by an equally electrifying violet shed out back. The house bears a sign in front that reads, TURTLE LOVER PARKING ONLY. VIOLATORS BETTER SHUT THE SHELL UP.

Parked in the drive are a white Smart car and a black Scion. Both are mounted with strobe lights—like the ambulances they are. Emblazoned with the Turtle Rescue League logo, and with stickers urging fellow motorists, STOP FOR TURTLES IN THE ROAD— HELP THEM GET ACROSS, the cars serve as emergency vehicles for transporting injured turtles to the thousand-square-foot turtle hospital that occupies the basement of the home.

Observed by closed-circuit TV—one of several security measures taken because even sick or injured, turtles are so valuable on the black market that the patients here could be targets of abduction—my friend, the wildlife artist Matt Patterson, and I mount the steps to the wooden deck and knock on the door. Alexxia Bell, Turtle Rescue League's president, lets us in. She's forty-six, slender, tall, and seems dressed for a party, wearing a black nylon long-sleeved shoulderless shirt and slim pale-blue velvet jeans. Once inside, Matt and I carefully step over a knee-high wooden barricade to enter the living room.

We are soon met with the reason for the barricade. Pizza Man, a twenty-year-old, twelve-pound red-footed tortoise, with a knobby black-and-yellow shell, is headed toward us like a slow-motion missile. High-stepping on columnar legs, his toenails tapping softly on the wooden floor, he holds his pale-yellow bottom shell, or plastron, tall as he paces determinedly across the room. He stops two inches from my feet. He pointedly jerks his wizened-looking head to the right, holds it still for a second, jerks his head back to the center, then jerks it to the left. He then swings his neck back to center and stares up into my face.

Such a spirited reaction from a turtle might come as a surprise. While most people like turtles, many, even biologists, for years dismissed the reptiles' intellect as little more sophisticated than a pet rock's. Compared with their sometimes colossal bodies (a leatherback sea turtle who washed up on a Welsh beach holds the record, at nine feet long, weighing over a ton), turtles' brains are remarkably small, which was believed a sign of low intelligence. "Turtles don't *need* intelligence," one field biologist, Alex Netherton, asserted in an online forum, "so they do not waste energy on it." Because turtles are famously slow, and spend considerable amounts of time stock-still, it's easy to get the impression they don't think or feel or know much—or do much of anything at all.

But clearly, Pizza Man is giving me a signal. It feels like a welcome.

"This turtle really loves attention," Alexxia explains. I bend down to stroke his soft, outstretched neck and head, admiring the red patches on his cheeks and nose and around his soulful, dark eyes. Then Pizza Man marches on to meet Matt. If anything, Pizza Man's ardor now grows. Even though it's February in New England, Matt, as usual, is wearing—along with his signature turtle-print headband—flip-flops, and Pizza Man makes a point of standing directly on the warm tops of Matt's feet as he performs his greeting.

Pizza Man's enthusiastic reception bodes well for us. Matt, already a renowned natural history artist at thirty-eight, and I have driven two hours from our homes in New Hampshire here to Southbridge, Massachusetts, to ask a favor. Since the previous summer, when we started helping friends protect a nesting site for five species of New England turtles, we've been drawn deeper and deeper into the world of these beloved but little-understood reptiles. Last year, we attended a turtle summit, a symposium for turtle rehabilitators, here at the League. We came away as stunned as if we'd visited Lourdes.

Alexxia had projected a slide of one of their patients, a female snapping turtle. The entire first third of her shell was shattered, three of her legs were smashed, one eye was gone; she had been lying on the side of the asphalt road where she'd been hit, cooking in the sun, for hours. But two years later, she was returned to the wild, healed. "What looks like a fatal injury to some animals may be survivable to a turtle," Alexxia had told the assembled crowd. "Basically, if the turtle's organs are not smeared all over the road, you might well be able to save her. We never give up on a turtle."

So today, we are back. We want to take part in this miracle. We've come asking if, once the busy spring season gets underway, we might be allowed to volunteer at the League's hospital, and help broken creatures be made whole.

Alexxia picks up Pizza Man and kisses him on the head. ("I won't get a tortoise disease," she insists. Though even healthy turtles may carry the *Salmonella* bacterium—one reason the sale of many baby turtles was banned—Alexxia tells us she's more likely to get an infection from kissing a child. ("I'm a hundred degrees," she explains, "and he's a reptile.") It's clear that Pizza Man is used to Alexxia's kisses; even though he has just been hoisted into the air by a mammal six times his size, rather than withdraw his head at her touch, he extends it.

"Pizza Man always wants to be wherever I am," Alexxia says. Among the more than 150 turtles currently in residence here— turtles recovering from illness or injury, turtles relinquished by previous owners and awaiting new homes, native baby turtles who hatched out late or too small and will be released in the spring, turtles who were born deformed or who are permanently disabled, and will live here forever—Pizza Man, rescued from the basement of a drug dealer, is one of the very few whom she considers a personal pet. A decision was made early on not to bond too deeply with the others, most of whom either were slated for release or, in the case of non-native and non-releasable turtles, might be adopted out to good homes elsewhere. But Pizza Man is an exception.

The other is Sprockets, a thirty-pound Burmese mountain tortoise. At age twelve, he's not even a third of the size he'll reach at maturity. He may live for one hundred years. With a dark, boxy head and prominent beak, he comes plodding out of the first-floor bathroom on legs as scaly as a Monterey pinecone. At the very same moment, Natasha Nowick, forty-four, sporting the TRL logo on her green polo shirt, with matching green streaks highlighting chin-length dark hair, descends from the office upstairs.

Sprockets's emergence as Natasha, cofounder of the League, enters the room is no coincidence. Sprockets is as devoted to Natasha as Pizza Man is to Alexxia. Belonging to a species native to Myanmar, Malaysia, Thailand, and Sumatra, Sprockets was found on a September day five years ago, wandering around the park next to Worcester Polytechnic Institute. His owner had dumped him. He was briefly taken in by some engineering students. One of them was Natasha's younger brother, who called Natasha and Alexxia to come pick the tortoise up. "He was so nervous," Natasha remembers. "Every inhale was a quaver, and he hid in a corner." But once Sprockets settled in at the League, he would fall asleep in bliss on Natasha's lap. And soon, she tells us, "he started telling us his life story, vocalizing and bobbing his head. He would vocalize for twenty minutes at a time."

Most of us think of turtles as being silent, but no: Some are quite talkative, and various species croak, squeak, belch, whine, and whistle. (For the Velociraptors' barking in *Jurassic Park*, filmmakers used the sound of tortoises having sex.) Some species of Australian and South American river turtle nestlings communicate vocally with each other, and with their mothers, while still inside the egg.

Natasha describes Sprockets's voice as "a grunt crossed with air being released from a party balloon." He's not as vocal these days, she tells us: "At first he was like a kindergartner going on and on about his favorite subjects," she says. "Now he is more mature and reserved." But he still bobs his head when he's excited—which he is doing right now, seemingly in acknowledgment of what may appear, to him, the great commotion of our arrival. "He's very excited to meet you," Natasha tells us.

Turtles have distinctive personalities and experience strong emotions, Alexxia and Natasha explain. But because they lack mammals' facial expressions, it is difficult for humans to see this. The ancestors of humans and turtles diverged some 310 million years ago, back before plants learned to flower, before the evolution of corals and their mighty reefs, and not long after our fishy forebears crawled out of the water onto land. Yet you can, with attention and practice, intuition and empathy, learn to read the sometimes subtle, sometimes alien signals of turtles.

"We get to know them," Alexxia explains. "The personalities start to shine. It's an unspoken communication. But it's real."

This is key, the two women say, to their extraordinary success in saving, and often releasing back to the wild, thousands of turtles who otherwise would have died—including many so badly injured that even veterinarians specializing in rehabilitating wildlife would have euthanized them.

It has taken Natasha and Alexxia over a decade to get to this point. The two met twenty-one years ago, at a fashion store where Natasha was the manager and Alexxia had applied for a job doing makeup. Though opposites in many ways—Alexxia a flashy extrovert who partied at Boston dance clubs till dawn, Natasha a soft-spoken introvert who enjoys video games and data—they both love animals. One of Alexxia's earliest memories growing up was watching her dad help a snapping turtle cross the street. When Natasha was little, her family took in orphaned and injured wildlife, including a raccoon, a woodchuck, and a seagull.

One spring day, while the couple was headed for a date hiking on a local trail system, they found a turtle on the road, crushed but still alive, in obvious agony. It was clearly mortally wounded. They knew of no vet who could help or euthanize the animal. Feeling helpless, Alexxia placed the turtle headfirst under the tire of her car and ran it over, killing it quickly so it wouldn't suffer. Today, in fact, this kind of "mechanical euthanization" is considered a humane option for turtles who have no hope of recovery. But for the couple, "it was devastating," says Natasha. The pain of that moment lingers for them still.

The next day, they saw another turtle—unharmed, but marooned on the dangerous tarmac of the cloverleaf interchange of two highways. They picked it up and released it in a pond. "We kept finding turtles," remembers Alexxia. "So instead of hiking, one day we said, 'Let's find turtles and help them cross.'"

They created posters and flyers telling others how to help, too. But they kept finding turtles hit by cars, run over by lawn mowers and hay mowers, chewed by dogs, or afflicted by disease from neglect or poor care from people who had bought them as pets or had taken them from the wild.

"With an injured turtle in front of us," Natasha tells us, "we didn't know where to go. If we brought every one we found to a wildlife clinic, it would overwhelm them." And, she reminds us, "in 2008, there weren't smartphones, and most Internet searches were unhelpful. We were looking for answers in desperation—and they weren't out there. How could we change the world for turtles?"

"At the time, we couldn't do basic shell repair," Alexxia says. "But we learned."

They learned from the veterinarians at Tufts Wildlife Clinic. They learned from the director of the Mass Audubon Wellfleet Bay Wildlife Sanctuary on Cape Cod, which rescues endangered sea turtles. They learned from the head veterinarian at the New England Aquarium. One of their most inspiring mentors was a wildlife rehabilitator they met at a turtle intensive seminar in New Jersey. While treating thousands of patients at the New York Center for Turtle Rehabilitation, Kathy Michell had herself survived cancer, multiple sclerosis, and then a cancer relapse with a survival prognosis of six months—which turned into a year, and then two, and then five years and counting. "She imparted tenacity," Natasha said. "She taught us not to give up."

The first rescue they took home was a pathetically undernourished yearling snapper they named Nibbles. Alexxia discovered him exiled to a plastic shoebox in a quarter inch of dirty water at a client's home, and persuaded the owner to relinquish him. The couple immediately spent two hundred dollars at their local pet store to properly house and feed him. Soon Alexxia and Natasha found themselves living with seventy-five rescued turtles and half a dozen thirty-gallon stock tanks crammed into an 860-square-foot, two-bedroom apartment in Webster, Massachusetts. The many filters, heat lamps, and full-spectrum lights their charges needed overexerted the rooms' electrical capacity, and they had to run an additional conduit line—which, fortunately, Alexxia (who now owns her own appliance repair business) knew how to install.

Friends who'd visit would ask, "But where do you girls *live*?" They slept under a kayak, essential for water rescues, suspended from ropes over the bed.

"When we started, we couldn't have imagined we'd one day have this place," Alexxia tells us. (They bought the house seven years ago, back when it was, according to Natasha, "the color of neglect.") "And this sanctuary," continues Alexxia, "two cars, a board of directors, and Michaela . . ."

Petite, blond Michaela Conder, eighteen, has joined our chat as we perch on stools around a tall table while Sprockets and Pizza Man plod beneath our feet. She's shy and doesn't say much, but her blue eyes and big smile radiate eagerness and energy. She's the only other paid employee at TRL, and handles communications and largely running the day-to-day operation of the turtle sanctuary and hospital. Michaela first became involved with the group at sixteen, when she was visiting an aunt from her home in Kansas; she moved to Rhode Island, where she lives with her grandmother, so she could work here. "When I look into the eyes of a turtle," she tells me, "it almost feels to me as if the entire universe is held within them. They have so much understanding and knowledge." Michaela also works part-time at a coffee shop—but each week, in addition to her paid hours at TRL, she drives the hour and a half from her grandmother's house to put in additional hours of unpaid work as a volunteer for the organization.

She's always wanted to do something meaningful with her young life, but what that was wasn't immediately evident—which is why she had put off going to college. But now, she's found it. "That's another thing turtles have given me," she says: "Purpose. The turtles give me a reason to get out of bed every morning."

It's becoming clear to Matt and me that for these three people, caring for turtles is more than a job, more than a charity: It's a sacred devotion. "When I work on the bench with the turtles, I'm glad my parts don't match them," admits Alexxia. "In a season or two, I'd be out of parts. My blood, my bones—I'd give it to them."

. . .

Why turtles? Natasha and Alexxia have, in their years together, rescued other creatures, from squirrels to salamanders (including a skunk they found on their way to pick up an injured turtle; once inside their little car, and with an hour left to travel, the animal, in extremis, sprayed its musk). What makes turtles so special?

I'd been thinking about this for some time. Turtles do have a big fan base. In some cases, this is literal: At New England Aquarium, by far the most popular of its tens of thousands of animals is Myrtle, a ninety-year-old, 550-pound green sea turtle who has lived there since 1970. She has her own Facebook page with more than seven thousand followers. Turtles star as heroes of stories, comics, and movies, from the Tortoise in Aesop's 2,600-year-old fable, to the Teenage Mutant Ninjas, to wise Crush and little Squirt in *Finding Nemo*. Turtles are popular subjects of artwork, collectibles, and toys; there is even a Turtle Splash breakfast cereal, which comes with a free baby sea turtle adoption kit.

Almost everyone has seen a turtle, and most people my age once lived with one. Every dime store in the United States carried inch-long baby red-eared sliders in the 1950s, '60s, and midway through the '70s—along with tiny round terrariums, each with a spiral ramp topped by a plastic palm tree. Unfortunately this was a completely inappropriate habitat for animals who should command a home range measured in square kilometers and live for fifty years. And their food was wrong, too: Most of us were sold ant eggs to feed them, when baby sliders really need a variety of insect and invertebrate prey, as well as vegetables and other plants, to eat.

But it was no wonder these doomed infant turtles were such popular pets. A baby turtle fits perfectly into a child's hand. (And also elsewhere—which was why some kids were getting salmonella, and the sale of turtles less than four inches long, the width of a child's open mouth, was banned in 1975.) Unlike most reptiles, turtles don't frighten us; they seldom bite, and they don't slither or scurry, but move slowly enough that we can watch them for a while as they charmingly carry their "house" on their back. When I was growing up in Virginia and New York and New Jersey, every kid had one—often several in succession, because most quickly died. (My parents rushed to replace them before I found out.) All of mine were named Miz Yellow Eyes.

Like me, Matt loved turtles from an early age. "I've been a turtle nerd my whole life," he will tell you proudly. His earliest memory is of heading out in a rowboat with his father, a biology teacher, to look for turtles when he was three. Later, he and his dad built fenced outdoor enclosures for turtles they would find and bring home. "We didn't know back then we were doing anything wrong," he explains, now fully aware that taking native animals from the wild is illegal. "I just loved them and wanted them with me so I could watch them."

He's never lost that boyish Huck Finn attitude, nor his love of outdoor adventure. He's a bit feral. It doesn't seem like a house or an office could hold him for long—and it never has. After graduating from art school, he worked a scant two and a half years as a product design illustrator in different offices, "always looking out the window, dreaming of somewhere else." One of the offices where he worked was by a river, so he'd bring his kayak so he could fish and look for turtles on his lunch break. But as soon as he could, he quit the office to work for himself, concentrating exclusively on wildlife art—creating images so vividly realistic that when I saw a photo of one of his turtle paintings that happened to have his hand in the frame, I thought the hand was fake and the turtle was alive.

Matt's always ready to head out in his flip-flops to wade into a river or stream or swamp. Any place that's good for turtles is good for Matt Patterson, and he will go anywhere, and do anything, to observe, to paint, or to help them.

He's combined his turtle expertise with college wrestling skills to wrangle monster snappers into his canoe for close-up views. He's traveled to the Spiny Desert of Madagascar with the conservation group Turtle Survival Alliance to meet critically endangered radiated tortoises in the wild. He haunts herpetology conferences and trade shows where turtles are prominently featured—where, to his wife's dismay, a common question strangers ask each other is "How many herps do you have?" ("Sounds like the symptoms of a horrible disease!" she remarked.)

He's been with his three-toed box turtle, Polly, longer than he's known his spouse. He's been married to Erin, a speech and language pathologist, for a decade. He's been with Polly for twenty-four years. The largest of his four pet turtles is Eddie, an African spur-thigh tortoise who he thought at first was male. She's only twenty pounds now, but she could grow to one hundred pounds and live up to 150 years. (He plans to build Eddie her own barn, and has provided for her in his will.)

What inspires such allegiance? Once Matt wrote out his feelings in an email to his mom. Though she tolerated with grace a home overrun with animals of all kinds—escaped snakes, a pet alligator, and a collection of turtles that once spanned fourteen species—to Matt's continuing dismay, her appreciation for turtles falls short of his own.

"Did you know," Matt wrote to her with the zeal of a disciple, "the first animals to orbit the moon were tortoises?" (They were a pair of unnamed steppe tortoises aboard the Soviet Zond 5 space probe in September 1968.) "Did you know some species live over two hundred years?

"Turtles are as old as the first dinosaurs, older than the first crocodiles, and have been around for over 250 million years. Unlike us," he explained, "turtles are extremely important to the planet's biodiversity! Some turtles, like snapping turtles, are the vultures of the ponds, lakes, and rivers, eating dead and decaying animals and plants. Gopher tortoises are considered a keystone species," he continued, noting that over 360 other species depend on this one kind of turtle, and their burrows, to survive. Other turtle species are equally essential to their ecosystems: Hawksbill sea turtles protect coral reefs by eating sponges, and other sea turtles eat jellyfish, which keeps them from overpopulating . . .

Matt pressed on, filling a full page of single-spaced type. "This is why I love turtles," he concluded, "and this is why my work is dedicated to helping protect them."

Many of these same reasons inspire Alexxia's, Natasha's, and my own love for turtles, too. Turtles' familiarity is a doorway to appreciating their strangeness. Turtles are unlikely, surprising animals. The more than 350 species of turtles, gracing six continents, display breathtaking talents. One, of course, is their longevity: A turtle who recently died at age 288 was alive when George Washington was born, when homes were lit by candles, medicine largely consisted of enemas and bloodletting, and mental illness was treated with powder made from the hooves of moose. Another turtle had a baby at age 140. Some can sense a lake or pond a mile away; others migrate across entire oceans to find the very beach where they hatched decades earlier. Some breathe through their butts; some pee through their mouths. Some stay active under ice-covered waters; others climb fences and trees. Some are red, some are yellow, and some change color dramatically once a year. There are turtles with soft shells, turtles with necks longer than their bodies, turtles with heads so big they can't retract them, turtles whose shells glow in the dark. Some turtles could even help us cure cancer. Asian mayapples, the source of etoposide, used to treat lung and testicular cancer, have been harvested almost to extinction. American mayapples provide an effective substitute, though the plant's seeds are extremely difficult to propagate—unless the seeds are ingested and excreted by box turtles.

Alexxia deeply respects turtles and their powers. But she also finds them hilarious. "They look so goofy," says Alexxia. "Let's design an animal that will persist for almost three hundred million years. You wouldn't design a turtle, with those pretty shells. You'd design something with big jaws, a huge brain—not something that falls upside down and can't get up."

Natasha concedes that "a lot of people might think it's completely ridiculous that we spend this amount of time and money and energy on turtles." True, most people like turtles, and many love them. "But how many times have I been standing at a booth at an event when someone asks, 'What is the purpose of turtles?'"

Alexxia gets frustrated when people ask what turtles have done for us. "They don't have to do anything for us!" she says, her temper flashing. "What are we doing for *them*?"

"Why turtles? Why art?" asks Natasha. "Why have children? Why anything?"

"They were here first!" Alexxia insists. "They are what life is, what life does—and they are worth saving."

"Here's an animal that walked with the dinosaurs," Natasha explains. "The earth warmed and cooled, warmed and cooled, and they are still here. But we are messing them up. Why shouldn't we be driven by a desire to set things right for them?"

For Alexxia too, it's as simple as this: "Turtles need more help than any other wildlife," she says. She's right: Turtles are the most imperiled major group of animals on earth. Like other wild animals, turtle populations shrink when houses, roads, and stores displace their homes. They suffer from pollution, climate change, and invasive species. Cars run them over. Dogs, raccoons, skunks, and otters chew them up. And on top of all this, there is a murderous, monstrous illegal trade in turtles—for their meat, for their eggs, for their shells, and for pets. "Helping any animal is a good thing," Alexxia had told the group who'd gathered for the turtle summit last year. "Saving a chipmunk is a good thing. But if you save a turtle, particularly a female turtle, she may be able to lay eggs for another hundred years. With each turtle you save," she said, "you are saving generations."

"So this," she tells Matt and me, "is where we play. Turtles are where we make the world better."

"Can we be part of that 'we'?" I ask awkwardly.

"Sure," she says, and Natasha nods her assent. "C'mon, let me show you downstairs."

2.

In Search of Turtle Time

Percy, a one-hundred-year-old three-toed box turtle

The minute we open the door to the basement, even before we descend the stairs, we're transported to another world. We're engulfed instantly in 75-degree warmth and the scent of hundreds of turtles and tens of thousands of gallons of water, conjuring the warm green funk of a quiet pond in summer.

At the bottom of the stairs, the first thing we see is the surgical theater: spotless aluminum exam and operating tables; a high-intensity light and a magnifying viewer; a Doppler ultrasound machine, which can estimate blood flow through the heart and blood vessels; storage areas for surgical instruments, bandages, vet wrap, and syringes; scales for weighing turtles large and small; blackboards listing meds and procedures scheduled for various patients; a fridge and freezer for storing foods and meds; a stacked washer/dryer; baskets full of clean, folded towels; a deep double sink.

But of course, we are most eager to meet the turtles themselves.

As we round a corner, Alexxia raises her voice above the hum of pumps and filters: "This is Sergeant Pockets," she tells us.

Basking under a heat lamp beside the ramp leading to his fifty-gallon pool, his long, wrinkled neck extended, Sergeant Pockets lacks the red "ear" patches that give his species its common name. He is an exceptionally dark-colored turtle. And I've never seen such a big red-eared slider. His nearly black shell is nine and a half inches long—enormous for a male (females grow larger). "He's over fifty years old," Natasha explains. He's named in honor of a police sergeant who ultimately helped close down the food market where the big turtle was being offered for sale at $3.47 a pound. "He was really sick," Alexxia tells us. "He had pneumonia and metabolic bone disease, and couldn't use his rear legs. He's a non-native species"—sliders are native to the South Central United States, not New England—"and he's a grumpy guy. He can't be released. He's here with us forever."

On shelves beside, above, and below the sergeant are the woodsy habitats for box turtles and the watery tanks with ramps for five other large red-eared sliders. Natasha introduces each by name: Razz, Walnut, Acorn, Speedy, Cherry, Sammy, Willow, Cottonwood . . . "If we can find them the right homes, they'll get adopted," she explains. "Or, they can stay with us forever."

On another shelf, opposite Sergeant Pockets, lives Percy. Enjoying a spacious habitat with a peat moss substrate, plastic plants (live ones would be quickly destroyed), several shelters, and a soaking tub, Percy has an exceptionally smooth, domed shell and piercing red eyes. He's a three-toed box turtle who is at least one hundred years old. Dr. Barbara Bonner, a veterinarian venerated for her turtle rescues, had found him at a Massachusetts pet store, very ill from living on a concrete slab surrounded by water—completely inappropriate for a woodland turtle. Alexxia lifts him from his habitat and sets him on the concrete floor. To our astonishment, like a windup toy, Percy instantly *runs* toward Michaela. As she scoots backwards in play, she can barely keep up with the advancing centenarian.

"Yes, he chases you," Natasha says. "He's a rema[...]
having made the century mark. And part of his g[...]
show he's the man here. He's still in his prime!"

Christopher Raxworthy, associate curator of herpe[...]
the American Museum of Natural History, would agree. "Turtles
don't really die of old age," he told a reporter. The major organs of
a hundred-year-old turtle, he said, are indistinguishable from those
of a teenager of the same species. It almost appears as if turtles can
stop time. Their hearts can cease beating for long periods without
damage. In species that hibernate (in reptiles, it's called brumation), turtles can survive buried in mud for months without taking
a breath. In fact, if it weren't for infection or injury, the curator said,
turtles might just live forever.

But in a landscape dominated by humans and their machines,
there is almost no escape from trauma. Proving this point is the turtle we meet next, Snowball, a female snapper weighing perhaps ten
pounds, with a large teardrop-shaped scar on the front third of her
top shell. In her small, shallow stock tank, she's so immobile that
she looks dead. Her head tilts markedly to the right. She came in
three summers ago. "She had been hit by a car and dragged. She was
handed to us by another rescue group, who couldn't figure out how
to help her," Alexxia explains. Her back foot was so badly mangled
that most vets would have amputated it. Alexxia removed three toes,
cleaned her wounds, replenished her fluids, repaired her shell. She
fought infection with injections of antibiotics. She fed Snowball by
inserting a tube down her throat and into her stomach.

But time itself is the only thing that might heal her head injury.
"Snowball has neurological issues; sometimes she just flips over
and over." One night, after she'd been in the League's care for six
months, Snowball flipped over in the water and drowned. "I put her
on the Doppler to check for a heartbeat and I'm getting nothing,"
Alexxia remembers. "So she's dead. What have I got to lose? I take
her out and put a tube down into her lungs and I'm breathing for

ner. The next thing I know, I'm getting a beep on the Doppler. I recovered her heartbeat. So, I'm like, Let's see if I can save this turtle."

In a few hours, Snowball opened her eyes. Later that day, she showed some weak movement in her toes.

"It took three months to get her back to where she was before she drowned," says Alexxia. "She's like forty percent there, mentally, now."

"I think she's making progress," says Natasha.

"Perhaps half a percent improvement each month," replies Alexxia. "She's a slow train crawling up a big hill."

In the tank next to Snowball is Chutney, a somewhat smaller but still impressive male snapper, who came in spring two years ago with a similar problem. The car that struck him wounded his top shell, broke his jaw, and concussed his brain. "He was a roller," said Alexxia. He kept rolling over onto his back in his hospital box, and each time he rolled—because snappers use their heads and necks to push against the ground to flip themselves back upright—Alexxia would have to reset his broken jaw. "Any other clinic would have euthanized him," Alexxia said. There was thought to be no hope for cases like Chutney.

Alexxia and Natasha tried taping him down so he couldn't roll. The tape didn't hold. They tried to weigh down the top shell, or carapace, but they couldn't safely put enough pressure on the wound. They had to find another way. So they came up with an ingenious solution: They slid the snapper inside a Tupperware pitcher just wide enough to accommodate his shell. "The handle acted as a kickstand so he couldn't roll," explains Natasha. And because the pitcher was transparent, Chutney could still see around him; when the world stopped spinning, he would know. They called their invention the Chutney Tube, and it kept him upright and safe for four months—until he didn't need it anymore. One day, perhaps this spring, Chutney will be released.

"A brain injury is something you can recover from," says Natasha. "But the turnaround is long."

. . .

Everything takes a long time for a turtle.

They live slowly. They breathe slowly. (In cold water, an olive ridley sea turtle can hold its breath for seven hours.) Their hearts beat slowly. (The heart of a red-eared slider can slow to one beat per minute.) During the turtle summit, we were astonished to learn how slowly the patients here react to drugs. Many analgesics are useless, because a painkiller that would work on a mammal in seconds or minutes could take hours or even days to take effect on a turtle.

Turtles also die slowly—so slowly that *The Turtle Hub*, a website advising turtle owners on proper care, includes a video titled "How to Tell If Your Turtle Is Dead." Turtles' bodies are so different from ours that we can't judge the difference between life and death by mammalian standards: A 1957 newspaper article recounts that the heart of an alligator snapping turtle caught by a college student in Mariana, Florida, kept beating for *five days* after the turtle was decapitated. In laboratory experiments, even when completely deprived of oxygen, the brains of sliders can continue to function for days. For reasons like this, at Turtle Rescue League, Alexxia and Natasha never declare a turtle dead until rigor mortis sets in and/ or they detect the smell of decomposition. Until then, because of turtles' amazing healing powers, there is always hope. "We never give up on a turtle," Alexxia repeats.

But, while they heal remarkably, turtles heal slowly. "It takes time, but that's what we can give them," says Natasha. "Time is what turtles have."

Time is another reason why I am drawn to turtles. Time, like consciousness, is one of philosophy's "hard" problems, a mystery with which great minds have struggled for centuries and which has always held my fascination. I've often viewed time as an adversary. As a young journalist writing for a newspaper with five editions, I worked fourteen-hour days, facing the pressure of beating five daily deadlines. During the thirty-five years after that, working as a free-lancer and author writing about animals, I relished my wild travels and creative freedom. There have been moments, in the company of a hawk, or our pig, or an octopus, I felt I could escape ordinary time. Matt experiences this, too: "When I'm doing art—I love painting—it's almost meditative. I get in the zone and time slows down. It's the same when I get out in nature. Time slows down and I don't think about these other things."

But for me, this break from the clock and the calendar is always too short. I travel a lot, and when I am researching new work, or on a book tour, or giving talks, sometimes it feels I am constantly leaping from my bed and rushing outside in my bathrobe in order to catch a plane. When I am actually writing the book or the article, I am always on deadline, the Sword of Damocles over my head. In my lucky and enviable adult life, I have wanted for nothing, save one thing: I have wished, always, that time would slow down.

Instead, time has done the opposite. "When we were young, time crawled along slowly," observes my best friend, the author Elizabeth Marshall Thomas, who's eighty-nine. It did for me, too: I remember, as a child, feeling that Christmas, or my birthday, or summer was so far away that I might not make it. When I was nine, it seemed like I would *never* be ten; at fourteen, sweet sixteen was a distant shore; and as a teenager I felt the years to adulthood stretching tortuously long. "But as we start aging," Liz observes, "time flies." In her book *Growing Old: Notes on Aging with Something Like Grace,* she explains why: In our first two decades, we learn to walk, talk, read, study, swim, ride a bike, drive a car. We graduate high school and perhaps college. Many of us marry, have children. In those twenty years, we increase our body weight twelvefold; we change from a helpless lump of need to an upright and autonomous adult.

In her first two decades of life, Liz writes, "I had dozens, maybe hundreds of important experiences that changed who I was and what I did." Contrast this with the decades that followed. During this time, most people are doing basically the same things: building a career, nurturing a family. In the last twenty years, Liz writes, one of the few changes she counts among life-altering events is a new understanding of the use of the comma.

I felt a profound change the year I turned sixty. I felt myself ready to join a new phase of life. "Aging involves constructing a new persona," says the nature and travel writer Edward Hoagland, a contemporary of my friend Liz, "as one did in adolescence." At sixty, I had entered the age group of Elder—a role in which I could pursue an objective markedly different from those from every previous decade of my adult life, and arguably one that is more morally compelling: seeking wisdom. Who better than turtles—ancient, unhurried, long-lived beings revered as icons of serenity and persistence—to show me the path to wisdom, and how to make my peace with time?

Fire Chief's 120-gallon tank stands in a corner. "He's one of my favorites," says Alexxia. She lifts the wire lid over the tank and drops in a whole unpeeled banana. A head that looks to me nearly as thick and long as my thigh lurches out of the water as Fire Chief lunges to seize it. He gulps it like a crocodile. This huge snapper is probably sixty to eighty years old, Alexxia tells us. He weighed forty-two pounds when he first arrived on October 8 two years ago.

Fire Chief used to summer in a pond beside a firehouse, where all the firemen knew him. Like many turtles, Fire Chief occupied one pond in summer and used a different one to hibernate in winter. The firemen saw him coming and going between his wintering grounds and his summer pond in spring and fall each year. But the two ponds were separated by a major road, and one day he was hit by a truck as he was migrating to his winter digs.

"We got a report of the accident, but the Good Samaritan who called it in couldn't stay with him," Alexxia explains. She and

Natasha rushed out with their kayak and nets. "The whole fire department came out to meet us, they were all so concerned." By that time, the injured turtle had managed to crawl back into the firehouse pond. Alexxia got in the kayak and, incredibly, located him in three feet of muddy water. But Fire Chief saw her coming and submerged, swimming for a deeper spot—so Alexxia dove into the 66-degree pond and came up with the giant snapper in her arms.

The truck impact left a hideous wound on his top shell, but worse, Fire Chief's spine was broken. His back legs were paralyzed. But this is one of the astonishing things about turtles: They can regenerate nerve tissue, even sometimes when the spinal cord is actually cut in half. "It might take three months," says Alexxia. "It might take five years. But it might heal." Now, Fire Chief seems to have recovered at least some movement in his rear legs.

He couldn't survive like this in the wild—at least not at this point in his recovery. But Alexxia and Natasha have seen worse heal completely. That same year, another large snapper came to them from North Andover, up north toward the coast. He summered in a mill pond surrounded by busy streets and apartment complexes. He may have been returning from his winter home, or out looking for a female, when he tumbled off a retaining wall onto hard pavement. Before he was rescued, he had waited, with a cracked carapace and broken shoulder girdle, for so long that flies had laid eggs that had hatched into maggots inside his mouth.

Alexxia and Natasha fixed his mouth, glued his shell, and repaired his shoulder girdle. They let him go that same fall.

Then, there was Gill, named after the Massachusetts town where he was from. Alexxia and Natasha met the people who'd found the turtle on the side of the road at a Cumberland Farms convenience store, where they had brought the gigantic snapping turtle in their truck. Alexxia estimated he might weigh forty or fifty pounds. "When I went to pick him up, it was like you go to pick up a rock and it's really Styrofoam. He was only thirteen pounds! He was a skeleton with a shell." Gill had a healed wound on his shell but was sloughing off patches of skin, and he smelled like he was dead. His back legs and tail weren't moving.

Gill was in such rough shape that Alexxia and Natasha thought saving him was beyond their skill set. They took him to the renowned Tufts Wildlife Clinic, part of the Cummings School for Veterinary Medicine. The vets there recommended euthanasia. But the couple wanted to give Gill a chance.

What was wrong with this snapper? Why was he sloughing skin? Alexxia pieced together his story, which she read in the scars of his healing shell: "The year before," she explains, "he had been hit by a car. He crawled across the road and found a patch of grass and sat there. For a year. With no food, no water, nothing. He managed to get into a situation that winter wouldn't kill him off. So, when a turtle starves, the body doesn't replace its cells. The body conserves everything. But when he finally got food, he could replace those cells"—hence the sloughing. "But by then," continues Alexxia, "the biome of his gut was messed up, so he couldn't get much nutrition. So we fed him whole foods. Entire fish, for example."

Gill's back legs started moving. He began to gain weight. He stopped sloughing skin. He began to smell better. Two years later, Alexxia phoned her colleagues at Tufts: "You know that turtle that was almost dead? I'm releasing him tomorrow."

So there's hope for Fire Chief. There's hope, too, for the two young spotted turtles, unnamed, who came in last year. These are gorgeous little animals, with jet-black shells spangled with small yellow dots. Once common throughout the Northeast, they are now listed as federally endangered, having lost half their population in a single turtle's lifetime. One is a female with a shell crack and a rear leg problem; the other is "cognitively blind." Her eyes look normal, but when she was about to be released after several weeks of rehab last year, it became clear that couldn't happen. Natasha recounts, "We were sitting on the banks of the wetlands with this turtle and she was staring outward—but not taking in the visual world. So we stomped back out of the woods and went home. But that," she insists, "doesn't rule out eventual release."

Nova is also blind. She sits in the shallow water of her tank, looking inert. She hatched here, from eggs from a mother who lived in a polluted pond. In addition to blindness, due to brain damage, her day-to-night cycle, according to Natasha and Alexxia, is now one week long. She sleeps when she is upside down in her dry hospital box. Today coincides with the start of her weeklong sleep. Alexxia plucks her from the water, and the turtle struggles, back feet swimming in the air and front legs swiping at her eyes. But then Alexxia deftly flips her over on her back, lays her in her warm, dry hospital box, and places a plush turtle stuffie on her plastron. Instantly, to our amazement, the turtle relaxes. She has lived here for seven years. It doesn't look likely that she will recover. But even if she does not, she can stay here, in safety and comfort, for the rest of her days.

"This spring, we're going to try to start Fire Chief on physical therapy," Natasha tells us. There's a large fenced Turtle Garden in the side yard where, with supervision, he can start exercising his back legs.

But for now, he waits—which turtles are good at doing. New England's wild native turtles are spending the winter brumating— buried in pond mud, hiding in holes, their hearts and breathing slowed, awaiting spring's awakening.

At the Turtle Rescue League, winter is the slow time working with these slow reptiles. Alexxia and Natasha don't brumate the turtles in their care; they prefer their charges stay awake so they can keep tabs on them. But the turtles do seem to slow down in winter. New turtles still trickle in; turtles who were disturbed during hibernation, turtles whose owners give them up for adoption, turtles whose problems are beyond the skills of other rehabbers. The first turtle of 2020 came in on January 8—the day another driver hit Alexxia's car at an intersection in Worcester. A neighbor who witnessed the accident noticed the car's logo from her window and came running out—to give up her unwanted Asian box turtle. They promptly named her Crash.

"May is when things really start up around here," Alexxia tells us. "It's just solid work for two months, sunup to sundown. And then in the morning, somebody finds a turtle on their way to work, when I've just spent the night picking maggots out of a turtle's body cavity. June is just banana-sandwich-nuts crazy, putting turtles together . . . "

It sounds like, come spring, they could use our help. We can hardly wait.

3.

The Turtle Crisis

A Rote Island snake-necked turtle

The many weeks until our volunteer stint at the Rescue League would begin in the spring seemed to yawn before us, nearly interminable and—except for spending time with Matt's Eddie, Polly, and his Russian and Hermann's tortoises, Ivan and Jimmy—distressingly low on turtles. But Matt had a solution: To further my turtle tutelage, to help me understand how varied, surprising, and imperiled these beings are, he arranged for a visit to his friends at the Turtle Survival Alliance's Turtle Survival Center outside Charleston, South Carolina.

This is one of the largest and most important breeding colonies for the world's most critically endangered turtles—some species of which, Matt explained, no longer exist in the wild. He had visited it as part of his very first turtle conference, in 2017. "It's wicked awesome," Matt told me. "You'll see turtles you won't see anywhere else in the world."

We flew out in March, on an abnormally clean and empty plane. We had heard about the spread of a new respiratory disease from

China; just the day before our flight, a cruise ship off the coast of California was held at sea when twenty-one of forty-six people tested were found to have the virus. But we weren't particularly worried. Matt had returned from Madagascar unscathed after eating, night after night, stews made from meat riddled with maggots; on my research trips in the tropics, I had survived numerous ailments, including dengue fever. How bad could this new contagion be? I was far more concerned about finding Matt's friend, Cris Hagen, the Center's director of animal management, at the airport.

But Cris was easy to spot. A serene, tall forty-seven-year-old, his gray hair parted in the middle, he was wearing an orange Turtle Survival Alliance shirt and gray cargo shorts. My attention was immediately captured by his left calf: It's covered with six large tattoos of the heads, in profile, of critically endangered South Asian and Southeast Asian river turtles of the genus *Batagur*. Each, he told us proudly, was modeled on a photograph of an individual Cris had personally met. He has more than fifty tattoos on his body, including Star Wars characters (he is known to some as the Yoda of turtles), droids, and a large radiation symbol (he used to do radioecology work in a lab). Lightsabers adorn the inside of his two index fingers, and a huge, extinct marine crustacean called a trilobite occupies one bicep. He mentioned that he even has one tattoo, the name of a favorite rock group, Slayer, inside his mouth.

"Who's going to get to enjoy that?" I asked. "Your dentist?"

"The tattoos are for me," he replied mildly. All the tattoos have meaning for him—which is why he is well on the way to executing his plan to have all fourteen living families of turtles inked on his skin. Turtles are a huge part of who Cris is.

On the fifty-minute trip from the airport to the Center, as we passed yellow-bellied sliders basking on pondside logs and on the cement banks of drainage ditches, Cris told us about how he came to wield the tremendous responsibility of safeguarding the future of some of the most imperiled animals in the world.

"I did not take a traditional path," he told us—one of his many comments that, we soon saw, proved Cris the Jedi master of understatement.

He grew up in Ohio—"a herpetological mecca," with twelve species of native turtles, five species of lizards, and twenty-five species of snakes. Cris was obsessed with turtles at age five and regularly attended herp clubs by age twelve—by which time he was already an accomplished public speaker, giving talks on environmental education and using amphibians and reptiles as his education tools. "I skipped a lot of school to hang out with scientists," he said. "I taught herpetology at a college before I had a high school diploma." (Actually, he never got a high school diploma; he eventually got a GED.) His story sounded rather like the familiar tale of the smart, nerdy kid who grew up to become a conservationist, scientist, or rehabilitator. But there was another layer to his life.

"I had a semi-rough childhood," Cris mentioned casually. His parents divorced when he was young. He ended up living with his dad, who was sometimes angry and physically abusive. Later we learned that Cris's youthful diversions, besides turtles, included a shocking litany of crimes. He did heavy drugs—his first tattoo, at age fourteen, was "LSD," which he applied himself with a sewing needle and India ink. He stole cars to wreck into other cars and to drive through the storefronts of local businesses. He led police and their dogs on prolonged chases through the woods. He once stole a forty-pound pig, slathered the creature with grease, and let it loose in a local department store. He set fire to two schools on three separate occasions. Why? "Just to see what would happen," Cris said calmly. "I was a mean, stupid person, but in a joyful way."

From ages eleven through seventeen, Cris bounced in and out of drug rehab, psychiatric hospitals, and juvenile detention centers. A probation officer was continuously assigned to him from ages thirteen to twenty-two. At eighteen, Cris spent much of a year in jail. "It was fun," he said, "doing a lot of drugs and playing a lot of cards. But you couldn't *go* anywhere." There were no turtles to be found in a psychiatric hospital, or in jail.

So he reformed his life. He spent much of his time volunteering with local natural history museums, herpetological societies, animal shelters, and the Miami Valley Serpentarium (now the Kentucky Reptile Zoo). In his twenties, he studied salamanders on Mount St. Helens. He monitored the nests of hawksbill sea turtles in Hawaii. He worked as a herpetologist for the Santa Barbara Zoo. He journeyed to Sarawak in Malaysia, intending to earn a master's degree, but couldn't find enough turtles for his project—so he kept traveling through Sulawesi and Komodo, looking for reptiles, and especially turtles, along the way.

He joined the staff of Savannah River Ecology Lab, a research arm of University of Georgia, in Aiken, South Carolina, in 2002, a post for which he traveled the world two to three months of the year. In his quest to find turtles, he has braved jihadists, earth-quakes, rockslides, and tropical diseases. In East Timor, he was, as he notes matter-of-factly, "caught up in some unrest that involved some gunfire." But it's been worth it: In several cases, he has been among the very small handful of Americans ever to observe certain rare turtles in the wild. He has coauthored the description of five taxa, or scientific classifications, of turtles new to science.

Cris has worked closely with Turtle Survival Alliance since its beginning. The group was founded in 2001 in response to what was termed the Asian turtle crisis. Just one example: In December of 2011, Hong Kong security police confiscated an illegal shipment of eleven thousand live turtles and tortoises of eleven species. What to do with them all? "A few thousand of them came here to the United States," Cris explained—the TSA had to find temporary homes for them all with zoos, universities, even backyard hobbyists. "Lots of people drove home with hundreds of turtles," he said.

In 2013, with the purchase of the fifty-acre property that, on a washboard road lined with small homes and trailers, we are now approaching, Cris began, as he says, living his dream job: managing one of the largest and rarest collections of turtles in the world, with an eye to one day replenishing lost populations in the wild.

"How do you handle caring for *six hundred* turtles?" I ask.

He *manages* the Center's collection, Cris clarifies. Other employees handle most of the physical, day-to-day husbandry of the turtles here. "That's why I have five hundred turtles at home," he tells us. "I *need* to take care of turtles."

The entry to the Turtle Survival Center is festooned with eye-catching signs: NO TRESPASSING. SNAKE CROSSING. BEWARE OF DOG. WARNING: MOVING GATE CAN CAUSE INJURY OR DEATH. Coils of razor wire top the eight-foot-tall perimeter fence, augmented by two strands of electric wire. Motion sensors, lights, alarms, and a team of vocal, vigilant Rottweilers guard the property. Sensors are set to detect any glass that is broken, and sirens wail if anything happens that is out of the ordinary. Three employees live on-site or next door, and Cris notes, "We all have a lot of firearms." Everyone who visits must disable the GPS on their mobile phones so the location of the Center remains secret.

These precautions are necessary because of the very reason the Center is here in the first place. Turtles are a red-hot commodity in the ruthless world of the illegal wildlife trade. Like the underground market for guns, drugs, and sex, turtle trafficking is networked, clandestine, and lucrative. A single Yunnan box turtle could command $200,000 on the black market. A Chinese three-striped box turtle, whose powdered plastron is rumored (incorrectly) to cure cancer, can fetch as much as $25,000. In many parts of Asia—where most of the stolen turtles turn up in phony elixirs (often claimed, due to turtles' longevity, to preserve youthful beauty in women or sexual potency in men), as tortoiseshell accessories like pens and bracelets, or sold as prestigious pets—more than three quarters of the native species are either threatened with extinction or already gone from their natural homes.

So many Asian turtles have been "vacuumed from the wild," as one TSA video puts it, that now the reptiles are being snatched from ponds, woodlands, and seas of the United States to be illegally shipped to feed this malignant market. Sea turtles, box turtles, spotted turtles, snapping turtles—no turtle is safe. Poachers mine scientific data, and scour books and local newspapers for clues on where to find them. One researcher in Ontario found that after one of her grad students published the location of her studies, seventy percent of the wood turtles in her study area mysteriously disappeared—leaving no skeleton or shell behind. A turtle rehabber whom Matt and I visited, who was caring for fewer than a dozen turtles, asked us not to reveal her name or town for fear that poachers would break in to steal her charges from her modest New England condo.

"Turtles are the most exploited and abused animals in the world," assert the authors of the authoritative *Turtles of the World*, Frank Bonin, Bernard Devaux, and Alain Duprès. This, they note, has long been the case: People have taken turtles and their eggs for food, for ornaments, and for medicine for millennia. But the pressures of today's illegal markets, driven largely by new Chinese wealth, are, turtle experts agree, without precedent.

The confiscation that inspired the founding of TSA was large, but not unique. In 2015, Philippine wildlife authorities confiscated 3,800 critically endangered Palawan forest turtles from a single wildlife trafficker—more of the species than were thought to remain in the wild. Within seventy-two hours of the discovery, TSA assembled a team of experts from three continents to feed, house, and heal them. In April 2018, ten thousand endangered radiated tortoises were seized from a single house in Madagascar, where they were packed like cobblestones, dehydrated and hungry, many with smashed shells and injured eyes. Again, TSA experts assembled keepers, vets, and construction crews to help. Six months later, authorities intercepted a shipment of seven thousand more.

"Turtles are struggling to persist in the modern world, and that fact is generally unrecognized or even ignored," writes the U.S. Geological Survey research ecologist Jeffrey Lovich and his co-authors in an article in the journal *Bioscience*. Turtles have "survived everything nature could throw at them from both Earth and outer space (e.g., the asteroid that wiped out the dinosaurs)," they continue, "but will they survive modern humans?"

The Turtle Survival Center exists to ensure the answer is yes. Its motto is "Zero Turtle Extinctions."

Cris opens the threatening metal gate, and as the guard dogs bark furiously, we walk past eight handsome, landscaped, rock-garden-lined pools for the Vietnamese pond turtles. There are believed to be fewer than fifty mature individuals of this small, dark species left in the wild. The ones here now lie hidden, brumating in mud and brush, awaiting spring's waking warmth. ("We're surrounded by sleeping turtles," Matt sighs dreamily.) They were successfully bred at the Center for the first time in 2013. Along with others hatched at zoos and captive colonies elsewhere, members of the species now number in the hundreds and are slated for repatriation into the wild.

Inside the main building, the first room we visit is dedicated to the youngest hatchlings. Layer upon layer of shelves are neatly lined with containers—clear five-gallon plastic tubs with circulating water, ten-gallon storage bins meant for sporting equipment, filled with peat moss or leaf litter—all bathed in the radiant heat and full-spectrum light from bulbs set on timers above, and rigged up to alarms should the lamps or power fail. Most of these tiny habitats sport "furniture" as well: ramps so the babies can choose to bask in favored 90-degree hotspots, little caves and underwater shelters for when they want privacy. Inside each container is a tiny living jewel, each an individual life, with value far beyond the measure of money. Some are among the most striking animals I have ever seen.

"It looks like a snake stuffed into a turtle's shell!" Matt exclaims as we bend over a clear plastic tray holding a two-inch-long baby. This hatchling's neck is longer than its shell. The neck so crowds his little tank that he must fold it into an S shape, like a serpent at rest. Because the neck is so long, it can't be retracted; if bothered, the turtle will tuck it to one side. This is a Rote Island snake-necked turtle, native to only twenty-seven square miles on a single island in Indonesia. The then-legal pet trade essentially wiped it out from the wild in just about two and a half decades in the 1970s through the 1990s. Only around a dozen wild-caught Rote Island snake-necks are known to remain in the United States' zoo population, where conservationists are breeding them with plans to transfer them to their home in the wild in Indonesia.

Near this little hatchling is an entirely different wonder: a spiny hill turtle. His four-inch-long, orangish shell is fringed with serrations all along the edges, and the crown of the carapace sports a tall ridge running its length. He reminds me of a Stegosaurus—but one whose bony plates have been hastily rearranged by a distracted god. "He looks like he'd be awful to swallow!" says Matt—and that is surely the evolutionary challenge these pointed edges address. Classified as endangered, it inhabits lowland and hill rainforests in parts of Myanmar, Thailand, Brunei, Malaysia, Singapore, Indonesia, and a couple of the southernmost islands in the Philippines. This baby, Cris says, just hatched in October.

Another habitat contains a two-inch Indochinese box turtle, with a rust-colored shell, red swirls and black swatches on the side of the head, and exceptionally large, beautiful, mesmerizing eyes. Each round, hole-black pupil is surrounded by a light-gold iris. His look like Cris's eyes, though the man's are silvery gray; they both remind me of the polished stones you find in a clear stream, and carry with them a hint of a stone's ancient patience. As my face hovers into its view, the little turtle focuses his eyes on mine. For a moment, it feels that the glance from this baby carries a knowing from an entire, timeless turtle lineage that can see at once into past, present, and future.

"Look!" Matt is bent over a black cement mixing tub filled with leaf litter. "It's a star!"

The shell on this two-inch baby is so steeply domed "he looks like a giant ladybug," Matt observes. But instead of orange wings with small black dots, the baby Burmese star tortoise boasts a banana-yellow shell covered with black geometric blotches. As the turtle ages, the black shapes will blossom, like the petals of a flower, to form the stunning starburst pattern that gives this tortoise both its name and its cachet in the illegal pet trade.

In the late 1990s and early 2000s, only a handful of these gorgeous tortoises were left in the wild; they were considered "ecologically extinct"—merely a relic, no longer a real player in the natural inter-relationships that help keep their ecosystem alive. But TSA helped turn that around.

Working in country with the Myanmar Ministry of Natural Resources and the Wildlife Conservation Society, TSA established three "assurance colonies" from 175 individuals, mostly confiscated from illegal wildlife traffickers, to hedge against extinction. In 2008, TSA took over the husbandry of the animals, and immediately the turtles started laying eggs—250 of them. As of February 2015, the consortium had bred 15,000 baby stars and released 2,100. Now they are breeding in the wild again.

"This is the modern-day equivalent," writes Steven Platt, a herpetologist with Wildlife Conservation Society, in an article in *Herpetological Review,* "of saving the bison from extinction."

The Burmese star tortoise's success story can be credited to many creative components. To protect the released turtles from poachers, their shells were tattooed with a message in Burmese—"Harm me, and harm comes to you"—which ruined the turtles' uses both as ornaments and as good-luck medicine. Released animals were blessed in ceremonies by Buddhist monks to stop locals from eating them. A number were tracked by radio telemetry.

But none of these genius efforts would have been of use without making more baby turtles to start with. How do you even keep such critically endangered turtles healthy, much less breed them, when nobody knows anything about them? That so few of these turtles are left makes the job both desperately urgent and titanically difficult.

On our tour, Cris shows us rarity after rarity: southern Vietnamese box turtle hatchlings, tiny beauties with yellow heads, their shells domed like a conquistador's helmet. The species was not recorded in the wild till 2010. He introduces us to Forsten's tortoises, with caramel-colored shells adorned with black splotches. In 2019, he and the turtle conservationist Christine Light were the first American researchers to ever lay eyes on this species in the wild. Before their sighting, only a single scientist had ever seen one outside of the marketplace or the pet trade. It is thought to survive only on Sulawesi and possibly one other island, Halmahera.

Our host leads us back outside, to a thirty-by-sixty-foot prefab greenhouse. In one of the many individual "apartments" of raised beds filled with plants, ponds, and caves, Cris plunges his arm deep into a pool of dark water, and like a magician pulling a rabbit out of a hat, comes up with a ten-inch turtle with a shockingly large armored head. "Here's another really cool turtle to show you," he says. This is an Asian big-headed turtle. Relative to the six-inch-long shell, his head is outsized—so big it can't be withdrawn. This is why members of this species often feel compelled to bite to defend themselves. The turtle gapes impressively.

"It's the last remaining representative of its family," Cris tells us—one of only four single-species families of turtles in the world today. "These are exceptional climbers," he continues. "We've had small ones climb cement walls as tall as my thigh." But they have had far more difficult problems to solve than thwarting potential escapes. "Only one zoo—Prospect Park, in Brooklyn—one private breeder, and us have ever bred this species in captivity in the U.S.," Cris announces.

It took Cris fourteen years to get these turtles to breed, lay fertile eggs, and successfully hatch the eggs. Today they lay eggs once a year. But first, Cris had to overcome a litany of obstacles. "They need to be cooled down at the right time of year, and require supervised breeding to avoid serious injuries. If the female is receptive, they breed in minutes. But then you have to take them out, or they bite each other, and the female can spend half a year healing from the mating. And," he added, "the female bites off the male's penis if given the opportunity."

Matt shudders.

"I can see how that would pose a problem," I comment.

But one after another, like Yoda wielding a lightsaber in slow motion, Cris is vanquishing the obstacles to his quest. As he leads us through the Tortoise Barn for the giant tortoises, to the greenhouse for the Sulawesi and Forsten's tortoises, through the Cuora Complex for the Chinese box turtles, Cris explains that for many species, breeding conditions need to be precisely right. Some successfully reproduce only once every few years. Rainfall, Cris explains, is often an important trigger. Some turtles need a cooling period in order for sperm and eggs to develop. Some require weird diets: Most species of turtles eat a variety of insects and plants, but some rely mainly on fungi, or need fresh bamboo shoots. There are others that are even more of an unknown, including their natural food sources, because they haven't been studied in the wild. "So much," says Cris, "is guesswork."

More often than not, Cris guesses right. In part, that's because he knows each of the roughly six hundred turtles in the Center personally. Some have lived in his house. And though he doesn't name them—he feels it is disrespectful to turtles who are not pets—he can recite each animal's history in savant-like detail.

This Pan's box turtle, with the green head? "She hatched at Zoo Atlanta in 2004, and she's lived with me ever since." She laid her first egg last year at age sixteen, but it was infertile. (Better luck, perhaps, this year.)

This adult Burmese mountain tortoise? "She's blind. I drove across the country to pick her up from a private turtle rehab and sanctuary in California. She weighs forty-five pounds. She has thirty-five eggs in her right now."

This adult Rote Island snake-neck? She came from Florida Zoological Imports in 1999, was purchased by a chemistry professor in South Carolina, then sold to a pet store. Cris met the owner on the side of a highway to buy her, so he could finally pair her with a mate. "That turtle had been nonreproductive for twenty years," Cris tells us. "In spring and summer of 2018, she laid two clutches of eggs. Neither hatched." A third clutch was laid in November. Most of the eggs cracked; one egg developed, but the embryo died in the shell. She laid a fourth clutch of nine eggs last November. All nine of those are still living—representing a unique genetic lineage from a lost population of wild ancestors. These will be donated to Wildlife Conservation Society for eventual release into the wild. Last fall, this same female laid more eggs—now developing in Center incubators.

But there's another crucial aspect to success in keeping and breeding turtles: "A good turtle person has to be able to put in the long haul," Cris says. "A turtle may spend ten years after you move them, just acclimating to the new place. Some species of turtles do nothing for half a year or more in the forest, just waiting for it to rain. A lot spend half the year just buried."

Turtles are supremely patient, and so is Cris. Turtle-kind has persisted through every imaginable kind of disruption: They have patiently survived through volcanic eruptions, ice ages, rising and falling seas, even the meteor impact. I imagine the Ancestor Turtle watching this all impassively, remaining calm in the face of calamity—rather like Cris watching the police and their dogs chase him through the woods, or the flames engulfing the schools he set on fire. He learned that life could be violent and chaotic as well as mysterious and joyful early on. "From about age ten," he had told us, "I've been able to face life around me. I just accepted it." And now? "I do what I do for the turtles as long as I'm alive, and that's it," he explained matter-of-factly. "Saving any species requires more than your lifetime," says Cris. "If people had more understanding of geologic time, we'd be less selfish and greedy, and think about the future."

It almost seems as if Cris is living in a different time-world than the rest of us. The rest of us humans, that is. At the moment of the world's most urgent extinction crises, he radiates calm. Amid a culture of hurry and hunger, he remains contented. He holds, at once, the in-the-moment present and the far-off future, and faces them both with composed persistence.

He is living in turtle time.

Turtle people are not your ordinary, run-of-the mill humans. Matt and I are staying at the TSA's intern guesthouse, an immaculate mobile home on a site where the former occupant raised rats for a living (and whose trailer was eventually relocated from the property). While Cris attends to a sick hatchling that evening, we hang out with keepers Clinton Doak, thirty-two, John Greene, forty-nine, and seven ornate box turtles brumating in a pen in back.

John and Clint came to work here by different paths. John, a big guy with a grizzly goatee and strikingly blue eyes, was once an actor and a model. Clint came up through the zoo world. But both men share with Cris an extraordinary passion. Clint once took out a five-thousand-dollar loan so he could take an unpaid position working with reptiles. He left his girlfriend, a lion and tiger keeper ("She always smelled like cat pee"), to move to Turtle Survival Center's headquarters. John never had any interest in starting a family. "If I had kids," he said, "I couldn't take care of so many turtles!" Both John and Clint agree with Cris, who was once married, but no more. "Relationships come and go," Cris told us, "but turtles are forever."

Working with turtles offers extraordinary rewards, they concur. It's gratifying to help a species that is struggling. It's interesting and edifying to get to know an animal that few other people understand. Turtles are beautiful and appealing creatures. But so are many others. Why turtles?

"There's something about the way a turtle looks at you," says Matt.

"I think it's the eyes," says Clint. Natasha, Alexxia, and Michaela had said the same thing.

"Turtle eyes are really special," agrees John. "Especially the *picturata*." (These southern Vietnamese box turtles are his favorites.)

"I see what you mean," I say. When a turtle looks at you, even briefly, it feels like more than a glance. "I think," I offer, "it's the intensity of their focus."

Such focus is rare in the human world. Attention is fragmented, our focus atomized. A study by UK's telecom regulator, Ofcom, found that their customers stopped whatever they were doing to check their phones on average every twelve minutes of the waking day; an earlier study found the typical American worker faced some kind of interruption every eight minutes. Over six million American children, by one count, are now diagnosed with a condition known as attention-deficit/hyperactivity disorder, with its symptoms of inability to concentrate, twitchy inattention, and lack of patience. The neuropsychologist Richard Restak, author of *The New Brain*, calls ADHD "the paradigmatic disorder of our times." But some experts now consider it not a disorder, but a type of brain organization adapted to this particular moment in human history. The best most of us can now muster may be "continuous partial attention," a phrase coined by ex–Apple and Microsoft consultant Linda Stone. We are alert enough to scan the world constantly but never concentrate on anything.

Earlier that day, at the Cuora Complex, Cris had introduced us to the Chinese yellow-headed box turtles. They were about to emerge from brumation. Just one was already awake: Her six-inch-long brown shell was submerged in her pond; her golden head was out of the shell, but still covered by water.

Her species is extinct in the wild, Cris told us; the last wild one known to be collected was bought at a market in 2013. "These animals are absolutely irreplaceable. There may be only one hundred wild-caught adults left on Earth," Cris said.

We were looking at one of them. But she was still not moving. "She can stay like that for hours," Cris reported. As we stared at her, enchanted, for ten minutes, Cris recited her particulars, including the names, addresses, and circumstances of the three people who kept her before she arrived at the Turtle Survival Center in 2013. The last guy lived in Arizona and gave her up after a wildfire threatened his private turtle holding facility.

"What a journey," Matt commented in awe.

And then:

"Nose up!"

"Eyes out!"

After our ten-minute wait, being rewarded by her attention felt like when the sun comes out from behind a cloud.

"Nobody pays attention like a turtle," I observe to our new friends that night.

"And nobody is as patient as a turtle," Matt adds.

In Arabic, the word for "patience" (*sabr*) comes from a root meaning "to confine or contain." Turtles, with their marvelous shells, literally embody the concept—they cannot, of course, come out of their shells, which are fused to their skeletons. The shell is one reason turtle-kind has persisted so long on this earth, and why they are blessed with such long lives. "Nobody can be given a better or greater blessing than patience," exhorts the Persian Islamic scholar Muhammad Al-Bukhari.

Later that night, Matt and I contemplate the years—no, decades—Cris has spent among thousands of slow reptiles, calmly offering them the gift of his patient attention. He has never been deflected from his goal, by either family or finances, or a quest for recognition, or a hunger for things. All that waiting . . . and he has never been bored. "I don't even understand how you can ever be bored," he'd commented to us at one point. "I can stare at a blank wall and have a great time. Boredom," he said, "is a human thing."

Yet he escapes it. How?

Matt and I both realize it at once.

"Because he's a turtle."

The next day, in his unhurried fashion, Cris shows us the rest of the Turtle Survival Center. We visit the medical facility and lab, with endoscopy equipment and monitors, microscopes, digital X-ray machine—and, since the center exists to further reproduction, all sorts of equipment geared for success in that arena, including an array of vibrators (they're to aid in sexing males and collecting semen, not for the females).

We tour the incubation lab. The place is not terribly busy at the moment: Right now its incubators hold only eight eggs, from seven species. In July, says Cris, all the incubators—some the size of large freezers—will be packed with eggs.

Egg development is a delicate process. Turning the eggs can suffocate an embryo. Just a few degrees' difference in heat determines whether eggs hatch into males or females; due to climate change, most green sea turtles in northern Australia are female. A few additional degrees and you have no turtles at all. (Studies of the

Mary River turtle show that freshwater turtles, too, are at risk from climate change: Hatchlings of this endangered species incubated under higher temperatures showed reduced swimming abilities and a preference for shallower waters—where more predators and less food are found.)

Some species take six months to hatch. And sometimes, Cris tells us, you have to manually hatch an egg—a process that demands a steady hand and nerves of steel. "You don't know if the turtle is dead, alive, or far enough along to open it. You open a tiny window using forceps. If you see a membrane or a vein, it's too early. Sometimes you can try to close an egg back up." Once Cris opened a Rote Island snake-neck egg too early; the unborn baby was dead. Or so he thought. The next day, it was crawling around. "Today, he's still with us!" he reports. Another testament to the resilience of turtles—and to the severity of modern human pressures on such tough customers.

With an hour to spare before we must leave for our flight, we still have a chance to see five hundred more turtles. We leave the Center to visit Cris's house.

With his nine rescue dogs—one is a mixed breed who was found at the dump with heartworm, a broken back leg, and detached retinas—Cris lives in a trailer that is, he admits, falling apart, eaten by mold, and overrun by insects. But his is a great place to keep turtles. In the front yard stand dozens of giant tubs, some the size of Jacuzzis, lined up neatly on top of a layer of springy pine needles, all with running water systems. Cris plunges one tattooed arm past a skim of duckweed and water hyacinth into the cool water. "C'mere, bud!" he says—and again, like a magician, he pulls out a turtle with a two-foot-long shell—an alligator snapping turtle. "This one came from the Allendale Cooter Fest," he tells us. The annual event features turtle races, people races, a parade, food (including cooked turtle meat), and crafts. This turtle was left behind at the end of the event, and turned in to the ecology lab where Cris worked before he switched to working full-time at TSA.

Next bin. Arm in, turtle out: "Here's a female alligator snapping turtle who was confiscated from a murder scene in Columbia, South Carolina . . . "

And on he goes. A yellow blotched map turtle—"I have fifty of them." A razorback musk turtle, his shell pitched as steeply as a tent. Two Burmese narrow-headed softshell turtles, with heads so pointed that they look like tails; Chinese and Vietnamese turtles; four-eyed turtles with spots like eyes on the top of the head . . . and then Cris takes us around to the side yard, where we enter an outbuilding filled with even more turtles, and a few crocodiles. (One, a four-year-old African dwarf crocodile, once crawled out of the habitat and was loose for a week.)

Some of the turtles are rescues, some with deformities or disease; some have cost him thousands of dollars. He knows each as an individual, each endlessly fascinating, each a source of joy and fulfillment, wonder and contentment. The satisfaction does not dim with time.

"I enjoy being with turtles as much today as when I was five years old," he says. To him "if it's good, it's always good. I have watched *Star Wars* five hundred times. It's like listening to a good album. I love 'Sweet Home Alabama.' I never get tired of a good song." It's the same with turtles.

He loves coming home from work. "I walk around my house after work drinking a beer and cleaning tanks," he says simply, "and it's one of the best parts of keeping turtles. It's when you interact with them."

I think of the clear greeting we received from Sprockets and Pizza Man at Turtle Rescue League, and the deep affection they obviously feel for Natasha and Alexxia. How would Cris describe his relationship with the five hundred turtles who live at his house?

"Turtles recognize you as the individual who brings them food," he explains. "They know me.

"But they don't have to do anything for me. It's enough," he insists, "that they exist."

We flew back home on a plane even emptier than the one on which we had come. But we were full of plans. The next day was Matt's birthday, and we were celebrating with a trip to the New England Aquarium. We talked about an expedition to Southeast Asia, home to so many of the endangered turtles Cris was breeding at the center. Matt has a friend in Florida who runs a tortoise sanctuary, and he was eager to introduce me to Spike, one of the large and particularly friendly Galápagos tortoises he and Erin had met there. In August, we planned to return to Charleston to attend the TSA's annual Freshwater Turtle Conservation Symposium.

The next day, the aquarium, like the airport, was strangely empty. The day after, it closed. A BioGen conference across the street, at a hotel where aquarium staff had gathered the night before for a meeting, had already infected one hundred people in Boston with the new virus. The World Health Organization declared Covid-19 a pandemic. Two days later, the president declared a national emergency.

For people around the world, life as we knew it, including our perception of time itself, was about to change profoundly. We had no idea for how long.

Snapping turtles nest in sandy soils.

4.

Superpowers

Demonstrating the Wheel Well Grip

We're all wearing masks when we meet again, in May, at Turtle Rescue League. Alexxia's is black, matching her tights and stretch top with shoulder cutouts; Natasha's is a green check, to go with her TRL polo. None of us is thrilled about face coverings, but at least, as I point out, Matt and I get to cover our face with herps: his with turtles, and mine with tree frogs.

So much has changed since we were last here. Nearly two million Americans have come down with Covid-19, and nearly 100,000 have died. In hindsight, Alexxia thinks that she caught the virus back in January—she had all the symptoms of flu, but with an extraordinary, lasting fatigue—and she may have given it to Natasha, too. The economy of Massachusetts, like much of the rest of the country, has been stalled for two months. Offices and manufacturing plants have closed. Retail outfits have been shuttered. Playgrounds, public pools, athletic fields, bars, casinos, gyms, museums . . . all closed.

But none of this has stopped the flow of turtles who need rescue.

While Pizza Man sleeps in the first-floor bathroom and Sprockets enjoys a spa soak in a bin in the kitchen, Matt and I step over the wooden turtle barricade to note a new tank. A hundred-gallon aquarium holds three enormous female red-eared sliders. At the sight of us, they swim eagerly to the surface to pull their yellow-and-green striped heads out to get a better look. I gawk back at them. I'm used to seeing them as dime-store babies, but the shells on each of these are at least nine inches long. All are curious and excited, lobbying for a treat.

"People just dumped them in the wild," Alexxia explains angrily. As they are native to the American Midwest, these released pets can survive in the wild, but they do so at the expense of the native turtles they displace, and are considered invasive species. These three can never be released.

Next to this tank, an adult male wood turtle, a species of special concern in Massachusetts, is basking beneath the heat lamp above the spacious artificial turf island arising from his eighty-gallon kidney-shaped pool. Wood turtles are considered exceptionally smart animals, and in experiments, learn mazes quickly, with mental mapping abilities similar to those of rats. Ralph, probably eighteen to twenty-four years old, was part of a confiscation of fifty-one illegally held, wild-born turtles who authorities turned over to the League a year and a half ago. He'd been held in a flat, featureless tank—"Flat surfaces don't exist in nature," as Alexxia notes—and as a result, he was the sickest of the bunch, suffering from an infection on his neck and pus-filled wounds, like bedsores, on his arms and legs. "We've finally gotten him healthy," says Alexxia. The state must match his genetics with a location before he can go free.

"Time to get down to business," Alexxia begins, as Michaela, masked in pink chevrons, joins us. "We need to talk about the protocols here for working with the weirdest and strangest situation. This is a unique year. Nobody's ever gone through this before."

We stand awkwardly, eyeing one another, six feet apart, as

Alexxia lists the new arrangements: Inside, we all wear masks. Matt, Michaela, and I can use the downstairs bathroom to wash, but avoid the kitchen sink. We can store food in the fridge. And we will frequently spray our hands with chlorhexidine—a blue antimicrobial liquid used to sanitize surgical instruments. (We would do that anyway. In the turtle hospital, everyone must wash and disinfect each time they handle a turtle or touch anything in its habitat, to prevent any possible spread of contagion.)

Then there's the matter of equipping us. Matt and I are now official interns—which means we could be called upon to rescue a turtle at any time. We are each issued plastic rescue boxes with locking lids drilled with air holes to keep ready in our cars. The nineteen-gallon bin can hold most large turtles, and the four-and-a-half-gallon tub is sufficient for the small ones. Each contains two towels and a bottle of water. The latter, along with a pair of plastic gloves, is for our use "in case of gore." (Alexxia, we notice, does not mince words.) She warns we must never rinse a wound, even if it's dirty, or we could spread infection. We should bring our own heavy gloves for handling defensive snapping turtles. "The snappers you meet in the wild are not going to want to be picked up," she notes. She will show us how to safely do this later today.

We move back outside to settle into Adirondack chairs, where we can sit six feet apart. Red-winged blackbirds call *"onk-a-ree,"* from a nearby wetland, lazy wasps float through the air trailing filamentous legs, and sunlight pours down on us like maple syrup. Spring, at least, in this strange year, has reassuringly returned. But this one, as Alexxia says, "will be different from anything we've ever experienced before."

In the past, folks who brought injured turtles to the rescue were invited inside and often given a tour of the facilities. Normally, in addition to hosting the annual turtle summit we'd attended last year, TRL also welcomes school groups, Girl Scouts, nature campers. But not this year. The summit was canceled, and visitors normally embraced will be reluctantly turned away. Outside the front door, two large lidded bins stand ready with signs instructing people to ring the bell when they drop off patients.

And there's another important change afoot. Alexxia's expression signals that she's about to issue a particularly important admonition. "As far as the tortoises go, I kiss Pizza Man—a lot. But in this point in time," she stresses, "the *only* persons allowed to kiss Pizza Man are me and Natasha."

Matt and I concede that we can live with that. But Michaela, even beneath her mask, looks stricken. Alexxia quickly adds, "But Michaela can kiss Peppi." Michaela's brow smooths with relief. His name short for Pepperoni—he's the same species as Pizza Man—Peppi is one of Michaela's favorites. She had come with Natasha to pick up Peppi, along with an elongated tortoise named Apricot, after a Samaritan noticed two scaly heads pop out of a handhold cutout for a box of copier paper left abandoned at a busy crosswalk at the end of last year. Michaela and Peppi had bonded instantly.

Since the start of the year, Alexxia tells us, twenty-six new turtles—animals who were relinquished or confiscated—have already trickled in to Turtle Rescue League. But that trickle will soon become a tsunami. "As of yesterday," announces Natasha, "the nesting wave entered the mid-Atlantic."

From our work with our friends protecting turtle nests for the two previous years, Matt and I know about the nesting wave: It's when native turtles are on the move. Emerging from their wintering grounds, they begin roaming widely, first males seeking mates, then mother turtles searching for sites to lay their eggs. This is when they are most likely to run into trouble.

For most of their life on the planet, adult turtles' marvelous shells have protected them from almost every predator. But humans have, in the last two centuries of our brief tenure on this planet—a tic, a blink, compared with the more than 250 million years of turtle time—turned these ancient creatures' world into a minefield. During the nesting wave, turtles, plodding along at an average of three miles an hour, have to cross roads crammed with four-thousand-pound vehicles whizzing at speeds that exceed fifty-five miles an hour. A survey by the State University of New York biologist James Gibbs estimated that in the Northeast, Great Lakes, and southeastern U.S., in areas crisscrossed by roads, up to twenty percent of the adult turtle population is killed by cars *each year*. A Florida State University researcher discovered that in his study area, near a busy road around Lake Jackson, only a quarter of the adult turtles in the population were female. He surmised that the missing quarter of the population—the animals who statistically should have been females—had fallen victim to wheels. Results of another study, concentrating on snappers living in an Ontario-area wetland bisected by a highway, were equally dire: Here, in the seventeen years between 1985 and 2002, the snapper population dropped from 941 to 177. The researchers predicted that things would only get worse. The snappers would soon be gone from the swamp.

Vehicles aren't the only human-caused hazards nesting mothers face. Even if they manage to cross the roads, dogs and cats chew them, lawn mowers and farm equipment shred them, curious children harass and kidnap them, and asphalt and concrete displace their nesting areas. This spring, turtles will need our help—and we have to be ready.

Via her network of turtle researchers and rehabilitators, Natasha has been monitoring the North American nesting wave since it started, in Florida, in February. "It picks up speed through Northern Virginia, then Delaware, Maryland, and Connecticut," she explains, "and then it's on us within days—and the floodgates open."

Much needs to be done before the wave hits. The incubators downstairs must be readied, heated, humidified, and filled with soil that has been commercially sterilized. The League will incubate the eggs from nests that were disturbed or dug in unsafe areas. They will incubate eggs from injured patients. Some turtles will "stress lay" in their hospital boxes; others, who have trouble expelling their eggs, can be induced with an injection of oxytocin, a drug similar to Pitocin, the synthetic hormone used in human maternity wards to hasten birth. And some eggs will be surgically removed from the bodies of patients who didn't make it. Even if a turtle is dead, Alexxia and Natasha can help.

To make space for new patients, healthy turtles must be cleared for release. With Michaela, Alexxia and Natasha review the candidates, marking their tanks with masking tape proclaiming RELEASE in red felt-tip. Now up for review are Chutney, the "roller" who nobody thought would make it; Hera the Terror, a feisty snapper who had been taken from the wild and raised in a Holyoke apartment, where an inappropriate diet made her top shell grow in the shape of an upturned bowl; a Blanding's turtle who was being raised at a school, and came in with an intestinal blockage last year; three musk turtle babies, who were hatched from eggs retrieved from a road-killed mother who had been dead for a week . . . and about seventy others, including dozens of undersized or late-hatching baby snappers and painted turtles who spent the winter under League care. When each patient is admitted, a thorough history is recorded on the hospital computer system, so that each turtle who is healed enough to be released can be returned to or near his or her original home in the wild.

Matt and I will assist with releases, traveling to wetlands all over Massachusetts. We'll often be needed to drive. For most days of the week, Alexxia is at work at her other business, fixing appliances, till afternoon. Michaela comes in as often as she can, but can't work every day. And Natasha can't drive, because she is blind.

Matt and I didn't know. Natasha navigates around the house like a sighted person. We'd noticed that behind the delicate, bejeweled glasses perched on a perfect nose above her high cheekbones, her gray-blue eyes never looked at us straight on; she only glanced at us from the side. We thought that was because, while Alexxia tended to be bold and blunt, Natasha was shy.

"I've not been out very long about being blind," she tells us. Others in her family also have retinitis pigmentosa, a genetic disorder that causes gradual loss of the light-sensitive tissue in the back of the eye, but the disease has progressed differently and more slowly for her. She has always been determined to use what vision she has for as long as she can. Even though her vision was failing, Natasha had first studied mechanical engineering in college, then switched to art photography. "I was always interested in visual things," she says. "One of my hobbies as I was going blind, believe it or not, was archery."

Even now, Natasha can see some things—but only pieces. She finds ways to fill in the blanks. "I have adaptive technology," she explains. Her cell phone recites her emails and texts, sped up because she can process hyperfast speech. She has a white cane for navigating outside the house, which she has named, affectionately, Mr. Stickey. With a screen magnifier, she can see ten percent of a computer screen. But since age twenty-one, she hasn't been able to see print on a page. "Being able to read a sign across the room seems like a superpower to me now," she says.

Happily, Natasha has superpowers of her own. She builds handsome turtle habitats and even fine furniture. She still plays video games—even with limited vision, she can marshal the relatively intact motion-detecting cells in her retina, and her sharp reflexes, to play some of the "slide and shoot" games—and she can rely on extrapolation and memory to enjoy role-playing games like Dungeons and Dragons. For exercise, she runs—with the help of a radio-controlled airplane wheel at the end of her cane. "I've learned," she says with a laugh, "to tuck and roll when I hammer into a curb." She's even thinking of resuming her pursuit of archery—after nesting season is over, of course.

Because of her blindness, Natasha has to do some things differently. When she reaches into a turtle's hospital box or tank, someone must tell her which is the front end of the turtle. "Once Alexxia forgot, and handed me a big red-eared slider headfirst," she says. "Chomp!" (Surprisingly, the big red-ears are far more likely to bite than the oft-maligned snappers, and Alexxia assures us that the bites "hurt like a sonofabitch.") This is one reason why the dashing, nimble Alexxia handles the surgeries and other emergencies at the League, and thoughtful, patient Natasha tends to the incubators and most day-to-day operations.

Blindness itself may enhance Natasha's other senses. Her sense of smell is exquisite. For instance, Blanding's turtles, she notes, "smell like Froot Loops" to her, a scent not even Matt can detect. When they're dry, snappers smell like buttered popcorn; wet, some smell "verdant, like cooked collard greens," while others carry the scent of an orange, and still others "border on spicy." She has another sensory talent, too: synesthesia. The word comes from Greek, and means "perceive together," for synesthetes experience one sense through another. For her, each number brings to mind a color: in her case, two appears in her head as blue, and nine as deep red. Six is a golden-butterscotch color, and evokes a butterscotch flavor— but it's also associated with migraines she had as a kid, so sixes annoy her. Four is no good either, she told me one day, but eight is OK. Only between one and four percent of humans have this ability, and it's thought to be a mostly happy accident of the wiring of the brain.

That blind people can develop their other senses more than sighted folk is borne out by research. A Massachusetts Eye and Ear study compared scans of the brains of sighted people with people who'd lost their sight before age three. They found the blind subjects had developed neural connections between nonvisual parts of the brain that were absent in sighted subjects. Another study, published in the *Journal of Neuroscience*, scanned blind and sighted people as they listened to various tones, and found the blind subjects better able to capture finely tuned frequencies of sound. Some blind people even learn to use hearing to navigate. Daniel Kish, who's been blind since he was thirteen months old, is sometimes known as the real-life Batman—because like a bat, he can "see" with sound. He issues clicks made with his tongue, and listens for their faint echoes as the sound waves bounce off objects in his environment. He says anyone can learn the technique, which he calls "flash sonar."

Sensory juggling occurs in other animals, too. A paper published in *PLOS Biology* detailed how even microscopic roundworms, experimentally deprived of the sense of touch, then became supersmellers, able to find food using far fainter scents than their peers. And some species of animals live just fine in the wild without sight: The Texas cave salamander, the southern cave crayfish, and the blind albino cave crab, for instance, all evolved from sighted ancestors, but they possess no eyes at all. Others, like the star-nosed mole, still have eyes but don't need them. As they swim through the moist soils of marshes and fields with their big pink hands, these beautiful five-inch animals rely on twenty-two pink tentacles, or rays, sprouting from the tip of the nose. There, twenty-five thousand supersensitive Eimer's organs allow them to use touch to navigate their dark world.

Natasha may indeed be using her other senses to fill in the details of what she can no longer see. Or she may be using still other parts of the brain. The Massachusetts Eye and Ear study also found blind people have increased connections in parts of the brain controlling memory, language, and sensory-motor function. The take-home message from that research, said one of its lead authors, Dr. Lofti Merabet, is that "the brain has tremendous potential to adapt."

Many of the turtles Natasha and Alexxia have known have shown this is true. One was a snapper they named Bazooka—because when they got her last year, it looked like she had been hit by an anti-tank rocket launcher. "That turtle looked really messed up," Alexxia remembers. Bazooka had been picked up by a kind-hearted woman, who was horrified by what she saw: The turtle's jaw was broken. Her shell was shattered. She was missing toes. She could probably see out of only one eye.

But when Alexxia examined the patient, she found these were *all old injuries*. They had already healed—probably years ago! The only thing wrong with her now was that she had a relatively minor crack in the shell. They released her back into the wild in a matter of weeks—where, "like a person with a disability," Natasha tells us, "she's almost certainly doing fine."

There are plenty of turtles in woods and wetlands, Natasha assures us, with healed wounds, with one eye, with a misaligned jaw or only three legs. And that doesn't stop them from living and loving their wild and precious lives.

"You're never seeing a turtle looking defeated," says Natasha with a smile. "It's such a joy to see their willingness to allow us to work with them."

Her words reveal yet another one of Natasha's superpowers: hope. And as we prepare to help heal the pain of so many innocent creatures, as we enter the uncertainty and fear of a global pandemic, we're all going to need plenty of it.

One of the important skills one needs to learn in turtle rescue is how to pick up and carry a large, unhappy snapping turtle.

These turtles are portrayed as vicious monsters. The author and adventurer Richard Coniff describes them as "big and scary, with dazzled, demented-looking eyes." The late Peter Prichard, a respected Florida turtle expert, once waved a broom handle in front of a 165-pound alligator snapping turtle, a relative of the common snapping turtle, whose range is confined to the American Southeast. With one chomp, the turtle bit the broomstick in half. Anecdotal reports abound of turtles biting off fingers and toes;

there is, in fact, one account, "Traumatic Amputation of a Finger from an Alligator Snapping Turtle Bite," published in the April 2016 issue of *Wilderness Environmental Medicine*. (In reality, Alexxia points out, "I could bite your finger off better than a snapping turtle." The jaw strength of a snapper, published in an article in *Journal of Evolutionary Biology*, measured between 208 and 226 newtons. The power of a bite from human molars? Between 300 and 700 newtons.) Up north, common snappers are maligned and feared almost as much as their much larger cousins; if they grew as big as alligator snappers, the east Texas herpetologist William Lamar told the author Stephen Harrigan, "they would take bathers regularly."

Snappers never intimidated Matt. He figured out how to handle them long ago. To get a better look at them, I've seen Matt grab huge, gaping snappers out of murky waters and hoist them into his kayak—without being bitten or even rocking the boat. Though they usually open their great mouths (which might well be a natural reflex), they seldom seem stressed enough to snap.

But for me, moving a big snapper has always been a big production.

One time, I was driving back from the airport, eager to get home to my husband, dog, pig, and chickens, when I noticed, in the opposite lane, a snapping turtle, her shell about a foot and a half long, emerging from the forest in an attempt to cross NH State Route 101.

What to do? She looked too heavy for me to lift, and I was afraid of being bitten. The only tool in my trunk of potential use was a turquoise collapsible umbrella. This I unfurled and placed, like a bright blue curtain, in front of her face. It served both to stop the turtle and to attract the attention of oncoming drivers not to run us both over.

But what now? The umbrella stopped her cold, but how could I get her across? All I knew to do was stay with her, holding the umbrella, until the traffic eventually stopped long enough for me to escort her across the road as she walked at her own pace.

I realized that I might be waiting till nightfall.

All those stupid drivers, I fumed, who don't even stop for a turtle on the road! What is wrong with these people? Can't they slow down for even a moment to save a life? Don't they know the turtles crossing the road are pregnant mothers, laying eggs to ensure the next generation? I began to curse them all under my breath.

And then the cars started pulling over.

"Need help?" A woman with two kids and their babysitter in the car waved and called from her window as she pulled over in the oncoming lane. Another woman also pulled over and got out to help. Then a third car stopped on the opposite side of the road. "I have a rake!" shouted a tall man as he emerged from his car. "Can you use it?"

As I held the turtle at bay with the umbrella, one of the women emerged from the forest with a fat stick. She presented it to the snapper, thinking the turtle would bite it and then we could pull her across. But the turtle, launching partway into the air to seize the stick, bit it in half. Then I had another idea. "Does anyone have a cardboard box?" I asked. "We can dismantle it and pull her across like a sled."

The first lady went back to her car and did one better. Thanks to her two kids, she happened to be carting around a plastic toboggan her children used to slide down hills, snowy or not. It came conveniently equipped with a long rope pull. We used the rake to coax the turtle onto the sled. At the first break in traffic, while the man watched for oncoming cars, I pulled her across. As the mother turtle hissed and snapped, we gently dumped her out of her sled, safe and sound. We all cheered.

That incident briefly restored my faith in our species. But I could have dispensed with all that drama had I known then about the Platter Lift.

From one of the stock tanks downstairs, Alexxia pulls out Number 96, a roughly twenty-five-pound female, rescued last year after a collision with a car, and carries her upstairs and out the door to demonstrate.

"She came in completely a zombie," she says. But once Alexxia lays the turtle on the grass of the front lawn, it's clear that Number 96 is zombified no more. As she sees me walking toward her, she wheels around and opens her impressive jaws. She is perfectly capable of lunging and biting if necessary. I step to the side, but she turns to face me again.

"See, as you approach the snapper, it turns to face you, and before you know it, you're doing the snapping turtle dance," Alexxia says. She's seen it often, with each partner hopping, lunging, and wheeling, both growing more fearful and jumpy at every turn. It looks hilarious, she says, especially when a burly, uniformed male cop is being held at bay by a relatively small turtle. That's when she loves to calmly pick the animal up, leaving the officer and any bystanders agog at her magic.

"So the first thing you want to do," Alexxia explains, "is go the long way around and approach the turtle from the back."

"Slowly," Natasha adds. "Gently. Calm your mind. The turtle can often sense your emotions. If you're fearful, they know. If you're calm, they often tune in to it."

I take a deep breath, wait a moment, and approach slowly, respectfully, lovingly from behind. Number 96 surely knows I am there. She can feel the vibration of my footsteps. Turtles do not have ears like ours—a plate of cartilage on each side of the head serves as their eardrum—but they can hear low frequencies, in the bass range, with great sensitivity, so she can hear me walking as well. And she can probably also smell me. But she doesn't whirl, and she doesn't snap.

"Now, take your hand and slide your palm under the plastron from the back," Alexxia instructs. You must never pick up a turtle by the tail, she warns—this can break the turtle's spine. But with the Platter Lift, it's good to use your other hand to hold the base, not the tip, of the tail, just to keep the turtle steady. "Now, just gently use your palm to lift her up."

She's right: It's just like a waiter holding a platter. Except the platter I'm holding is a wild reptile that scares most people too much to even approach.

"See, your hands are nowhere near her face. She can't bite you." She doesn't even try. She does drop her lower jaw in a gape—what Natasha and Alexxia call the snapping turtle smile—but neither one of us feels threatened. After I take a few steps and set her down, she doesn't lurch or spin, but stays calmly in place.

Another technique is called the Wheel Well Grip. This works for particularly heavy snappers. "It's good to be wearing gloves for this one," advises Natasha. In this position, the turtle can't bite, either—but if the turtle's back legs are "air swimming," the claws can leave a nasty scratch. "And," she adds, because the still waters in which snapping turtles live are seething with life, "it almost always becomes infected."

In this grip, explains Alexxia, you place the four fingers of each hand underneath the back of the top shell into a surprisingly capacious hollow between the carapace and the back legs—a spot like the wheel well on a car. With thumbs atop the carapace, you hold the turtle so the head is up and the tail is down, its back facing your front. In this position, the plastron faces outward, like you are holding a shield.

I find this a very secure grip indeed. I feel like I could carry Number 96 like this for quite a while. Delighted, I pose, holding a very calm, impressively large snapper, while Matt takes a photo.

But handling snapping turtles, Alexxia reminds us, is not the hard part of what we're doing. "I want everyone to understand," says Alexxia, "that you're going on some of our missions, and sometimes it gets a little hairy around here. Tempers can flare up," she warns.

This I can well imagine. Some of the situations we might face could move any animal-lover to rage.

"And sometimes, the turtles just break your heart," Alexxia tells us. "I always have a deep, deep feeling for the turtles. If I could take their pain for them, I would."

Caring deeply comes at a cost. The word *compassion* contains within itself its emotional price. The prefix *com* means "with"; the Latin root *pati* (from which we get the word *passion* as in the *Passion of Christ*) means "suffering." To feel compassion, then, is to enter into another's suffering, to suffer along with them. Carl Frederick Buechner, theologian and author, defines compassion as "the sometimes fatal capacity for feeling what it is like to live inside someone else's skin. It is the knowledge that there can never be any peace and joy for me until there is finally peace and joy for you."

In the 1980s and 1990s, psychologists finally coined a term for what the traumatologist Charles Finely called "the cost of caring": compassion fatigue. It's a complex of symptoms including exhaustion, irritability, and anger resulting from caring for others who are afflicted by fear and suffering. Also known as "secondhand shock," compassion fatigue is the bane of wildlife rehabilitators and veterinarians—as well as doctors and nurses coping with emergencies like battlefields, natural disasters, and pandemics. Compassion fatigue can exact a terrible physical toll—including headache, sleeplessness, weight loss, and chronic exhaustion—but worst of all is what Carla Joinson first noted was happening to her overextended fellow nurses: losing their "ability to nurture."

This is what worries Alexxia most of all. "Emotionally, it takes me . . . Last year, I fought it real hard. I thought I had unlimited capacity. I had to understand I am a human person—a human woman. I had to engage in other things."

To help herself unwind, Alexxia writes poetry. She rides motorcycles and dirt bikes. She explores abandoned factories, mills, old houses. And she loves dressing up. She makes it a point to wear makeup, do her hair, and put on a pretty outfit every day.

For Natasha, her release is exercise. Walking, running—and this year, she's ordering a recumbent bike called a tadpole, which will be custom-built for her needs.

The two are always looking out for each other during the frantic, tragic spring. They make a point of planning date nights—and if their date gets canceled, as it often does, by an emergency, they'll try to take a break for a lunch at the clam shack, or go to their favorite ice cream stand for a cone and a chance to listen to classic rock played on a tinny PA speaker in the parking lot. On tough days, Alexxia will stop by the donut shop to grab a coffee and treat for Natasha, even though Alexxia herself is a fan of neither. When Natasha sees Alexxia getting overwhelmed, she'll insist her partner take a breather on the back deck, and will handle the emergency herself with Michaela's help.

Still, it's hard. "By the end of the season, I'm a basket case," admits Alexxia. "I have only a couple strings holding me together.

"We are going to lose patients," she warns us. "Sometimes we're saving lives. Sometimes we're just moving a turtle out of danger. Sometimes we're working with people, and changing hearts and minds. There will be great times—and sad times."

And there will be moments so strange, and others so wondrous, that we cannot even begin to imagine them.

5.

Time's Arrow

A painted turtle crossing the road

We've just returned from our first morning's chore as interns—the happy task of moving and reassembling a turtle habitat Natasha had built for a nature center's captive sliders—when the call comes in from the local animal control officer: Big snapper hit on Ashland Avenue. It's a busy street, just off a state highway, and only a few miles away.

Within minutes, the officer arrives with the victim in a pet carrier. Alexxia takes one look and pronounces, "This is an emergency situation."

We rush downstairs.

Alexxia and Michaela wash hands, pull on surgical gloves. Using the Platter Lift, Alexxia hoists the patient from the carrier to a clean pink towel on the shiny exam table. "Hello, little monster," she says to him. The turtle cranes his neck upward, slowly, agonized. "You're still alive . . ."

The snapper's long tail shows we have a male. A bloody, two-inch wound through a bony plate, or scute, on the lower half of the upper shell reveals the gleaming white of a rib. Normally, Alexxia says, an injury like this would not be difficult to fix. But this is only the most obvious injury, and isn't the real problem. "I'm worried," she says. "The car tire has gone across the whole turtle. The shell is depressed." She gently pulls on one of the mighty rear feet. The leg flaps lifelessly.

"His spine is broken," she says. "He's just a bowl of jelly back there."

The snapper's intestines, kidneys, liver, bladder, colon—all the organs in the lower half of his body—have been squashed by the vehicle that ran over him. His crushed ribs have become "a bag of razor blades for his organs," Alexxia says. After two decades of successful life in the wild, in as little as two seconds, he was smashed, then squashed, and finally, punctured.

With her index finger, Alexxia lifts the snapper's beak to look inside his mouth. "Just checking your glottis," she says soothingly. A turtle's glottis is a vertical slit through the back of the tongue, the opening to the respiratory system. "He's breathing," Alexxia says to us. "OK, baby pie," she says to him, gently allowing the mighty jaw to close. He does not attempt to bite.

With a syringe, she squirts Lactated Ringer's sterile saline solution into the wound, which at least is fairly clean, and too fresh to have attracted flies. She repositions the turtle on the table and braces the broken edges of the damaged scute with silver tape and superglue. A dozen leeches adhere to his wrinkled neck, and these she removes with surgical tweezers, dropping them into an empty tub that once held margarine. "I've seen leeches drill right through a turtle's shell," Natasha says quietly. A check of the plastron reveals another dozen leeches, which Alexxia also removes. The parasites are the least of his worries, but this extra act of tenderness nearly undoes me.

"We give everybody a chance here," Natasha says to Matt and me. "But be prepared. This is a bad prognosis. I've never seen a turtle go from this to alive."

"We're going to set him in a good anatomical position, and provide fluids and analgesics," says Alexxia. "Sometimes there's great comfort in just setting the body right. I can't do much more than this."

To calculate the correct dose of pain reliever, the turtle must be weighed. He's nineteen and a half pounds, confirming he is at least twenty years old—a little older than Michaela. Like her, he should be just starting out on a long adult life. And like many people her age, he may have been out on this fine spring day, hoping to find a mate.

Alexxia fills a large syringe with about four tablespoons of Ringer's solution from one of the bags hanging by the exam table. Michaela retrieves a glass vial of meloxicam, an anti-inflammatory pain reliever, from the freezer, and warms it under her arm. "He's obviously in pain," says Alexxia, "so he'll get the maximum dose of analgesic." But because of a turtle's slow metabolism, a drug that might bring a human relief in seconds may not take effect for hours. This and an antibiotic will be mixed with the hydrating Ringer's, so he will only need a single shot instead of three—another mercy.

Michaela lines a bin with a fresh towel and places him gently inside. She covers him with a second clean towel and closes the lid. She marks his number—34—and his weight on masking tape across the lid. "He'll get another sixty milliliters of Ringer's before the day's over," explains Alexxia. But for now, he will rest, safe and quiet, in the calming dark.

"Will you name him?" I ask, weakly grasping for hope.

"They go at least seventy-two hours before we give them a name," answers Alexxia. A patient is most likely to die within the first three days after an injury.

"But after that, we almost always give them a name. It connects you emotionally to the healing process."

But for me and Matt, that's already happened.

. . .

We check on the other patients who arrived in last week—the ones who are still alive. Quite a number have already died, or were DOA. The situation is grim. Number 32, a small snapper, came in yesterday. "His injuries were severe," says Natasha. He was nearly split in two, from the neck to the tail. Alexxia superglued the crack in the shell. He isn't moving, so they check for a heartbeat with the Doppler. There is none. Later his body will be bagged and stored in the freezer for burial in the fall.

Number 33 is an adult female painted turtle who came in three days ago. She's survived long enough to get a name: Tacos. (Puckish Alexxia likes to name snappers after snack foods; Natasha prefers more evocative monikers.) "She was a doggie chew," explains Alexxia. The dog's teeth gnawed off the back of her shell and into the tender skin beneath. But when Michaela lifts her, Tacos is lively, air-swimming and craning her neck. "Oh, you're so cute!" says Michaela. Tacos's liveliness may not indicate she's feeling better. "She's with eggs," Natasha announces. "A mother turtle is just overwhelmed with the need to find a place to lay. She is desperate to get that job done." But with an open wound so near her cloaca, the opening through which her eggs must pass, she can't be allowed access to soil to dig a hole for her precious load. Once Tacos is stable, laying will be induced with shot of oxytocin. Michaela changes her towels, just like a nurse would change the bedsheets, and she gives a shot of Ringer's in the soft spot behind a rear leg to hydrate her.

Number 27 is another severe case—a large adult snapper who was hit by a car and dragged. They're calling him Skidplate. He came in with most of his plastron missing. His bones were showing, his wounds infested with maggots. Natasha holds him in the Wheel Well Grip so Alexxia can examine the five enormous, mostly circular wounds on his plastron. His left side took a worse hit than the right, but his right front foot is injured as well as a huge area near the tail. Alexxia slathers the wounds with a fresh application of silver sulfadiazine, which acts as a sterile, protective layer between the body and environment—and costs seventy dollars for a fourteen-ounce jar. Skidplate squirms—but does not snap. "I know, kid," croons Natasha. He came in fighting and snapping. Now he's patiently tolerating being lifted out of his hospital box for two shots every day, for fluids, antibiotics, and pain relievers. "But even though he's a pincushion, he knows we're helping him at this stage of the game," she says. After his exam and needle, Michaela packs him into his bin tightly, with extra towels, almost as if he were being prepared for shipping. This is so he has no pressure points on his wounds. Alexxia says cautiously, "He actually could survive."

Last, we check on a nearly fifty-pound snapper named Chunky Chip, who's probably one hundred years old. He came in two days ago. He was here last year, too—with the same issue. He's well known in his Marblehead neighborhood, Natasha explains, and so loved that the locals feed him bananas. His problem? Alexxia holds up what she first removed from this ancient turtle's huge mouth a year ago: a massive, two-inch fishhook that went through Chunky's soft palate and emerged at the side of his eye. This year, she removed a smaller bait hook—one that might have eventually dissolved—from his jaw. Upon examining him she could see where yet another hook—one that had somehow come dislodged before he came in to rescue—had left a slit across his throat that was seriously infected.

"Chunky Chip represents a struggle," Natasha tells us. "He has neighbors who love him so much that they feed him. And that means when he sees humans in a fishing boat, he's coming over, looking for a treat." The smaller hook might have come from kids who didn't know any better. "But the larger hook was not kids'," Natasha says. "That was a poacher's hook. And he's coming to it."

Chunky has probably lived in that pond for a century. It's not just his home; it's his world. But it's no longer his sanctuary.

So many of us humans feel the same way. We're not safe in our hometowns. We're not safe in our houses. Nowhere is safe. Germs that could kill us might enter our bodies at the post office, at the grocery store, with a handshake or a hug, with a package delivery, on the close, warm breath of friends and family.

May 28 dawned with the announcement that U.S. deaths from Covid-19 had surpassed 100,000. The next day, Matt and I learned that Number 34 had faded overnight. Already it was clear that counting deaths was the new way to mark time in a pandemic. Otherwise, it feels as if the arrow of time has simply halted in mid-air, going nowhere.

A popular cartoon shows a calendar with Monday, Tuesday, Wednesday, Thursday crossed out. Each square on the calendar reads only "Day. Day. Day. Day." Everyone seems to be living the movie *Groundhog Day*, in which the main character, played by Bill Murray, wakes each morning only to relive February 2 over and over and over. For weeks, spring seemed stuck, winter too stubborn to depart. We'd get a warm day, and then it would snow again. The daffodils seemed poised to dive back into the earth before the buds had a chance to appear.

Now, finally, with one long rain, the trees explode into leaf. The lilacs flower for Memorial Day—but for the first time since the 1860s, there are no parades on our small town's Main Street. Instead, in the cities, revolt. It appears that time has not stopped, but is actually running backwards, to an era before the civil rights achievements of the 1960s and '70s. Protests, looting, and shootings erupt in rage after an unarmed Black man spent more than nine minutes pinned under the knee of a white police officer, begging, until his death, "I can't breathe." The press widely reports that the president orders crowds tear-gassed so he can hold a photo op showing him clutching a Bible in front of a church he does not attend. "Congrats to the U.S. for having such inherent racism that it overtook a pandemic," one friend emails a colleague. "How proud to be an American." No wonder even summer doesn't want to come to us.

Cities and wildfires blaze, the virus rages, innocents die. But the violence and contagion are countered by acts of courage, kindness, and compassion. As volunteers repair smashed windows and torched buildings, as doctors and nurses repair stricken bodies, Matt and I, too, witness the work of healing the wounds of the world.

As we arrive at the League the next Tuesday, Alexxia is opening the mail. An envelope has arrived from a stranger. This happens all the time. Once again, a well-intentioned person she has never met has sent her a dozen or so hook-and-eye closures from old bras.

She's used to this. "I never thought I'd have a job where people sent me parts of their worn-out underwear," Alexxia mumbles. Several years ago, when the science of turtle rehab was younger, veterinarians and volunteers had success cementing these undergarment closures to opposite sides of a cracked shell, then using wire to pull the two halves together. Now Alexxia uses foil tape attached with superglue, which is neater and faster, and it doesn't catch on objects overhead like hooks and eyes would. But kindhearted people keep finding Turtle Rescue League on the Internet, and the bra bits keep coming.

"The craziness we get involved in!" Alexxia says. People call with turtle emergencies that sound unreal. Sometimes they are. Sometimes the pair will drive for hours to rescue injured turtles who turn out to be rocks or shreds of tires, or a turtle so dead that it's only a shell over a skeleton. Once they got a call that a turtle was stuck in an in-ground pool at a house that was for sale and unoccupied. "It was an animal death trap," Alexxia explains. "So we rushed out, and we managed to get a gas-powered pump and a fire hose, to go and drain a pool we don't own . . . and we didn't find any turtles." They did remove a drowning hazard, though, and built a ramp so any animals who fell in could get out.

Now, craziness has struck again. "Today, I gotta let you know," says Alexxia, "I got a report of a turtle with an arrow stuck in it."

My jaw drops.

Who shoots a turtle with an *arrow*? The studied meanness of it is shocking. So is the juxtaposition: the arrow, the symbol of speed that has come to stand for relentless directionality of time—piercing the body of a turtle, the embodiment of gentle slowness, of wisdom, of stability. What is happening to our world?

"I am not kidding. It's supposed to be at a pond two hours' drive away. I've got my volunteer Mike going out there to take a look."

It all sounds so impossible. How can anyone find a single turtle somewhere in a wetland complex two hours away, hours or days after it was last spotted? Could the turtle really have an *arrow* in it?

"We get calls like this," she says wearily. "A turtle is injured. It's somewhere in Massachusetts, somewhere in this woods or that pond. I've gone out days after a report of an injured turtle by the side of the road and searched the woods and found the turtle."

Alexxia sighs. "One day we'll get a call about a turtle climbing a skyscraper."

It'll be hours before we hear back from the volunteer. Meanwhile, Matt, Natasha, and I are dispatched to the Teamsters Union Local 170 in Worcester.

Throughout the spring, a parade of mother turtles march out of the neighboring swamp into the Teamsters parking lot to lay their eggs on islands of wood-chip mulch and lawn surrounded by asphalt. The large brick building and its two-acre parking lot were built fifteen years ago on what was, for countless generations, traditional nesting territory for native painted and snapping turtles. Now, trucks and cars pull in all day, putting traveling mother turtles at risk. Later, the emerging hatchlings will face even greater dangers, as they are harder for drivers to see. And in between, the eggs are vulnerable to predators like skunks and raccoons, for whom a turtle's nest is a protein bonanza. We're stopping by to check for nests, and to dig up and incubate any eggs in mortal peril.

TURTLE SEASON IS HERE! announce printed flyers taped to orange cones all over the parking lot. The man who created the signs, Scott Marrino, called the League earlier this morning. Scott, a fit fifty-four-year-old in a crew cut, is in charge of plumbing maintenance and groundskeeping at the Union, and ever since he was hired a year and a half ago, he's printed these signs, erected cones, and worked with the League to alert employees and visitors to the nesting turtles. "I come out here three or four times a day, and sometimes I find ten nests," he tells us. "My bosses don't pay me to patrol for turtle nests—but everyone around here does care about the turtles." Ever since he's been putting up the signs, he told us, "people are amazed. Everybody's involved here. Everybody's excited. The people who work here are mothers and fathers, too. And just to know that people care makes you feel really good."

Already this season, Scott has left wire screens, held in place with five-inch nails, over the disturbed soil he's found in four locations. The screens mark the suspected nest locations and also provide some protection against digging predators till we arrive, he figures. We'll check them all to see if eggs are present, and patrol the periphery of the parking lot for other nests in case they need protection or removal. "You're great people, doing this kind of stuff!" Scott says as we move off to do our job.

"We typically do get a lot of nesters here," says Natasha, navigating the parking lot with her white cane. We easily find Scott's four marked areas and begin to dig. For this, we all rely on touch—at which Natasha excels, but I am a bumbling neophyte. We dig without tools, using only ungloved fingertips, and gently—otherwise, we could break an egg. But if we're careful, Natasha advises, we'll notice the moment that fingers push through the ceiling of the nest chamber into the hole itself, and feel the cool, smooth curve of the uppermost eggs. "Tonight," she warns, "your fingertips will be really sore."

None of Scott's marked areas contain eggs, but it's quite likely that mothers dug here. Both paints and snappers dig test nests, and a number of turtle species—scientists have documented this among paints and two species of marine turtles so far—even dig decoy nests to fool predators.

Next to a retaining wall at the edge of the parking lot, Matt spots what looks like a fresh nest. But whoever was digging here left no eggs. We move on to other areas where we find disturbed soil. "The two pitfalls in this job are rushing—and giving up too early," Natasha counsels. Sometimes an egg chamber may be more than six inches deep. Patient persistence is essential. And you never know what you might find. Once, while digging at the edge of this very parking lot, Natasha uncovered not eggs, but a mother turtle. The tired mother had paused her digging and "decided just to rest in the cool damp." In fact, Natasha discovered, this mother had been digging in the same hole another turtle mother had excavated, laid in, and refilled.

Today, we find no eggs, and the turtle egg transporter—a plastic bin with a handle, filled with egg crate foam—will go back empty. "Nothing for today," Natasha reports to Scott, "but it's clear the turtles are out, and know the lay, as it were, of the land. Anything new, just give us a call!"

"Yes, I will," Scott replies with a wave. "Thank you for all you do!" he calls cheerfully. "I love your being here to help them!"

Back in the car, Natasha takes a call over the Bluetooth. She announces the news: "Mike Henry has found the turtle!"

Back at the League, Mike pulls up behind my car minutes after we reach the driveway. A slender guy in his late thirties wearing glasses and a trim dark beard emerges from the vehicle with the turtle in a forty-gallon tub.

"The animal control officer met me in the parking lot and gave me a net," he explains. How on earth did he find it? "I was lucky. It was a big turtle in a little pond." The pond wasn't actually that little—about eight acres. But Mike, who coaches software developers and facilitates their work, saves his vacation time for turtle rescue, and he's one of what Alexxia calls "our Super Volunteers"—a man of exceptional turtle talents.

In fact, Mike was the guy who brought in Pizza Man, having extracted the tortoise from a delicate circumstance when he managed to convince a potentially dangerous drug dealer to give up his pet turtle. Mike approached the situation with extraordinary diplomacy. "I heard you had a turtle who might be sick," he told the criminal. Pizza Man, then named Sparky, indeed had a serious respiratory disease, and was crawling around on the floor while the owner's large dog sniffed him. Mike has also worked on turtle restoration projects with endangered Blanding's turtles with the Massachusetts Division of Fisheries and Wildlife, and has adopted several turtles who needed homes. "Any time there's a turtle in trouble," Alexxia had told Matt and me earlier, "we can count on Mike to take care of it."

"I was going over a semi-submerged log when I saw him," Mike continues. "I'd never dealt with one in the water before!" After his first pass with the net, the water was muddied and he could hardly see. So he abandoned the net and plunged in with his hands. It took three tries to pull the giant snapper out of the water. "I started with the end I was really hoping was the tail," he tells us, "then Wheel-Welled him up and into the box. He didn't even try to bite me. He's a real chill guy."

He then carried the roughly thirty-pound turtle in the box for more than a quarter mile along the gravel path to his car, and drove the two hours from the pond in Marblehead to Southbridge.

Mike lifts the lid of the black plastic tub to reveal the patient. The snapper fills the space. And indeed, sticking out of the right side of his neck, angling in front of his face, and hitting his eyebrow and nose each time he extends his neck, is five inches of the shaft of an arrow. We dub the turtle Robin Hood.

"He looks like he has been living a good life," says Mike, patting the fat on the snapper's front leg. "He's a real porker!" He lifts the turtle up for us to examine. The plastron is smooth and clean. When he opens his mouth in the snapping turtle smile, it's healthy and pink inside.

All this bodes well for Robin Hood's recovery. But while we wait for Alexxia to get home from an appliance call, we need to try to research how she might get the arrow out.

A big concern is what kind of tip is embedded in the turtle's flesh. "We'll have to decide if it's even safe to remove," says Natasha. What's at the end of the shaft? A straight point? Or one with free-standing blades? "Some arrows are spring-loaded," says Matt. "We have to hope it's some fifteen-year-old idiot with a hobby-grade arrow," comments Natasha.

Mike again holds up Robin Hood in the Wheel Well Grip so we can get a closer look.

"It's a crossbow bolt!" cries Natasha.

Crossbows were originally developed as medieval military weapons, strong enough to pierce chain mail armor. Though today the terms *arrow* and *bolt* are often used interchangeably, technically a crossbow bolt is shorter and heavier than a traditional arrow.

"Let's see what Cabela's says for crossbow bolt," says Matt. A fishing, hunting, and outdoors outfitter, Cabela's has an online catalog, and surely, Matt reasons, we should be able to find this bolt featured there to figure out what kind of point lies buried in Robin Hood's body.

Matt finds it on his cell phone. "Six-point-three-inch aluminum," he reads. "No tip at the end. And it says it's 'high quality' . . ."

"Good to know," says Natasha with steely fury.

Who would shoot a turtle with an arrow? And such a *nice* turtle, too! "He's just a big baby," Mike croons. "He doesn't wanna bite me. You're a good man, pal," he says to the turtle as he sets him back down. "We're on your side."

Alexxia arrives and we all move downstairs. "Let's look at you!" she says as she Wheel-Wells him up to the exam table and flips on the magnifying light. "Hi, honey. Looka your big ugly face. You so ugly, you pretty!"

She reaches for the gold-colored shaft, which is inches from his jaws, and gently wiggles it. Robin Hood doesn't even gape. It's clear this is an old injury that has healed. The head of the bolt is embedded in his shoulder girdle and neck like barbed wire embedded in a tree.

Alexxia injects some painkilling lidocaine directly into the site. It might take fifteen minutes or more to take effect. She continues to feel around Robin Hood's neck and shoulder.

"It doesn't seem to have anything but a point on the head," Natasha informs her.

"Yes, I can feel it through the skin on the other side. But," Alexxia says, pulling on the bolt, "it's really in there." She pulls again, then twists. The arrow's not going anywhere. And though the big snapper isn't lunging or gaping, I can't imagine he's enjoying this.

"You so cute!" Alexxia says to her patient.

"Good man," encourages Mike. "Hang tight!"

Alexxia tugs and twists. Still the projectile doesn't budge. "I'm going to try to lubricate it with some Lactated Ringer's to inflate the area, so it might slide out." Another injection, and this time, she's able to slide a quarter inch of the shaft out. But there's still at least another inch of metal in his flesh.

She pulls some more. She tries twisting the bolt. Both her hands are right next to the snapper's face, one pressing on his neck, the other pulling on the shaft—and finally, the arrow yields. It's out!

"How does that feel, pal?" Mike asks the turtle.

"That's got to be such a relief," says Matt.

For a moment, I cover my face with my hands. I can feel myself breathe again.

Ironically, many human cultures, including Native Americans

who left their mark on ancient Southwest petroglyphs, have considered the arrow a symbol of life itself, so essential was this tool for their traditional lifeways. In the late 1920s, the arrow became the physicists' symbol for the flight of time. To Robin Hood, it was merely a source of pain and annoyance; to Alexxia, the weapon she's pulled from this turtle is merely more disgusting evidence of senseless human cruelty. But later, I will note the arrow is also considered a powerful symbol of struggle and triumph—and nothing could more aptly symbolize this moment.

Alexxia isn't done with her patient. "You got yourself a hat?" Tenderly she removes part of a dead maple leaf from Robin Hood's massive head. She notices what looks like a tiny blockage in one nostril. She attempts to remove it with surgical tweezers—and now, for the first time, the big snapper has had enough. He lunges and snaps. I jump nearly out of my skin—but this doesn't bother Alexxia at all. ("On the bench, I can feel the tension building. There's a lot that happens before a snap. You have milliseconds to prepare," she told me, as if that is a lot of time.) She anticipated it all, and simply moved out of the way. What looked like a blockage, she sees, is just a fleck of soil, and doesn't need messing with anyway.

She flushes the old wound with disinfecting Betadine, which pools like dark blood. She examines Robin Hood's face carefully. His nose is pink, like a fresh sunburn. There's a red sore on the right side of the neck, and an irritation over the left eye where the arrow's shaft had been rubbing against his face. She also notes a small scrape on the right front foot. All get rinsed with Ringer's and squirted with Betadine. Finally, she sets the turtle down on the floor. He heaves up his back end in self-defense, presenting the sharp, serrated part of his shell, an instinctual warning. "He may not even know he's doing that," says Natasha. Just for the record, we weigh our patient: 29.3 pounds. Even if he's a hundred years old, he's still just middle-aged for a snapper—relatively more youthful than I am.

"You're looking good!" says Mike to the turtle. And to Alexxia: "What's the plan now?"

Alexxia addresses the snapper. "You're a good guy. Now just relax in your box." Gently she replaces him in his carrier before turning to us. "His injuries are minor. There's no infection. His wounds are clean. Mike," she tells her volunteer, "you can release this guy today!"

Mike heads off with Robin Hood to release him back to his home pond in Marblehead, and Matt and I prepare to drive back to New Hampshire. When will we be needed next?

Turtle activity depends on the weather. The nesting wave is heading toward us. The forecast is warm and getting warmer; shortly the mercury could hit eighty.

"Thursday, I think it's going to be loco," Alexxia predicts.

"Hot and heavy," agrees Natasha. "And if we're not moving patients, we'll need your help doing releases."

We'll be back in two days.

6.

Beyond the Guardrails

A musk turtle lounges in a tree

The woodcock's whirring flight at dusk, the crimson pinpricks of maple buds, the sleigh bell chorus of spring peeper frogs—each familiar, beloved sign of spring's arrival feels like a promise renewed. Spring comes to us, ready or not—it comes through our windows, with the waterfall song of the wren at dawn; it falls on our lawns and our cars, like the catkins of willows and the yellow pollen of pines. Unfolding overnight like the petals of trillium and trout lily, spring opens its throat and shouts.

But a turtle nest is different. It's a whispered secret. I'd wandered woods and wetlands for more than thirty years before I ever saw one. Encountering the nest of a wild native turtle feels like pulling back a curtain on an unexpectedly tender moment, one that only a few are lucky enough to glimpse. To witness this feels quiet, intimate, and hallowed, a moment that stays with you forever.

My first time, I was with my friend David Carroll, an author, artist, and turtle sage. Long before he won a MacArthur Genius grant at age sixty-four in 2006, David would sometimes take me with him to the place he loved most in the world: a mosaic of swamp, marsh, meadow, swale, and river he called the Digs. Here, like a pilgrim, he would go as often as he could—daily during spring and summer—to meet the turtles.

David would park his car at the edge of an old logging road and we'd walk through a rusted gate that had been opened years ago and never closed—for, as he wrote in his first book, *The Year of the Turtle*, "there is nothing on the other side of this dirt road to close this gate against."

We were walking in a sunny, sandy area, hummocked and hollowed with frost heaves, full of bluestem grass, leathery bracken fern, and a stunted-looking species of goldenrod, with stems that seldom grow more than six inches tall. It was September then, and wild cranberries lay on the ground like small red marbles. All these plant species indicate what we humans consider "poor" soils. "But this is excellent for nesting turtles," David told me. Turtle nests need full sun. Dense herbs or shrubs would block the rays. "They don't even nest among the small gray birches and white meadowsweet," he explained. "That's too shady—"

And right then, something caught his eye.

It was a small opening in the soil, under two inches in diameter, and easy to miss among the tufts of grass and fern and the two-toed hoofprints of deer. But there it was: the exit hole, from which turtle hatchlings had dug their way from their nest, leaving their eggshells behind. David bent down and gently scraped aside some of the sand with one finger. He stood up, and to my astonishment, in his hand held a single, perfect, inch-long infant eastern painted turtle. The yellow yolk sac from the egg was still attached to the belly.

The baby, he said, was probably three days old. This turtle's siblings had already left the nest for the nearby wetlands. But most painted turtle hatchlings, he told me, choose instead to stay buried in their nest all winter, to emerge the following spring.

"This place is to me, hallowed ground," David said. Every time I accompanied him to the Digs, I felt that, too—as if we had passed through a magic portal to a different place and time. Here was a secret, sacred world where shy animals could go about their private lives as they had since the time of the dinosaurs.

Later, Matt took me to some of his favorite turtle spots, too. They could only be reached by boat. On a sunny day in early spring, still so cold I shivered in my down vest, Matt, in shorts and bare feet, took me to the place he calls Turtle Cove. He'd been going there for fifteen years. We put in our kayaks—he in his "Green Manatee," which he's had since high school, bearing scars from the teeth of an alligator he had met in Florida, and I in Erin's newer "Blue Blast"—at the edge of a gravel parking lot by an abandoned church. To Matt, an atheist, the church was only a landmark; to him, it was the water that was holy.

He led the way in the Manatee. Within minutes, we passed through a small space between the blueberry bushes, like a hidden gate—and found ourselves in the cove's embrace. We stilled our paddles. Beneath us, the cool, clear water held us like a cupped hand; above, the cheerful, spinning calls of the newly arrived red-winged blackbirds, the bouncing, chippy notes of swamp sparrows, the baying honk of geese flying overhead. Conditions for spotting turtles were perfect: It was the first sunny day in weeks. The buds of the water lilies looked like rockets poised to launch, but hadn't yet obscured the surface. The water was still and clear. Matt knew that turtles would be out basking.

We hadn't been in the cove a full minute before Matt spotted a four-and-a-half-inch-long painted turtle floating at the water's surface, his sides mottled with bright red, his neck striped with vibrant yellow. The turtle caught sight of us and dove, but Matt, his reflexes honed by a lifetime of catching critters, snatched him so we could get a better look. We could tell the paint was male from his soft, half-inch-long front claws, which he uses to seductively stroke a female's cheeks. The turtle didn't struggle. "He seems really content," Matt said before releasing him to swim under our boats.

Miracles happen in these remote, wild places. "Let's see if we can find a stinkpot," Matt suggested. They're also known as musk turtles; when startled, in an effort to ward off predators, they emit a yellow fluid from a gland between the front legs and the plastron. It smells, according to different reports, like stinky armpits or like burning electrical wire. But most people never encounter these little, five- to six-inch-long reptiles; their top shell is cryptic, dark brown or black. And further, most people don't realize that one of the best places to see them is up a *tree*. They often fall asleep there, sunning. And indeed, this is how Matt found the first musk turtle I ever saw.

Even more amazing was a different time, at the same place. We had just entered Turtle Cove when I saw a small dark shape with a yellow stripe inches from the water's surface, right beside my kayak. I plunged my hand into the water and, to my astonishment, pulled out a musk turtle. "Their dispositions," insisted Ann Haven Morgan in her 1930 *Handbook of Ponds and Streams*, "are consistently vicious. They stretch out their heads very slowly before snapping and the expression of the whole performance is surly and premeditated." But this miraculous little turtle, who had practically swum into my hand, was so calm and friendly that we couldn't even induce it to stink.

These secret turtle places recalled to me the refrain of Charles Baudelaire's 1857 poem "L'invitation au Voyage": *"Là, tout n'est qu'ordre et beauté / Luxe, calme et volupté."* ("There—nothing but order and beauty / Luxurious abundance, calm and sensuousness.") To the French poet's feeling of order and beauty, to his sense of calm so lush, it feels voluptuous, I would add, in describing how it feels to be in these wild places, a profound gratitude and humility: You feel privileged and humbled to glimpse these private lives.

"Turtles are shy creatures," notes Michael W. Klemens, director of the International Union for the Conservation of Nature's turtle conservation program, "wisely concealing their lives from our unwelcome scrutiny." And he is right, of course.

But it is also true, as I was to learn, that these ancient, secretive creatures are performing the most important and intimate rituals of their lives among us, hidden in plain sight.

When we joined her on that hot afternoon on the last day of May, Emily Murray, a sixty-five-year-old retired science teacher, was dressed in faded overalls, covered in bug bites and Band-Aids, and carrying more stuff—trowel, bucket, hoe, notebook, backpack, binoculars, and some unwieldy cylinders made of wire—than it seemed her slight frame could possibly handle. But her excitement was electric. "With a little luck," she told Matt and me, "we'll find the first nest of the season!"

Four of us—tall, blond Jeanne Richards, forty-eight, a part-time librarian and mom of four, along with Emily, Matt, and me—were standing just outside the fenced patio and pool in Jeanne's half-acre suburban backyard in Torrington, New Hampshire. Just a few steps more and we would pass through a wooden gate to cross into a flat, sandy area next to a baseball field.

It didn't look like hallowed ground. It was certainly not pristine; it was the sort of place you expect to find (and we did) plastic bits of a child's toy, a dog's lost ball, a rusted-out metal drum, and an old tire. A real estate developer would call this a "vacant lot."

But here, where we could hear dogs barking, kids playing, and homeowners mowing their lawns, no fewer than five species of native turtles emerge from surrounding waters and forests each spring to dig their nests and lay their eggs. Three kinds of these turtles—Blanding's, with shells domed high like army helmets; wood turtles, with bright orange spots on their arms and necks; and spotted turtles, their dark shells spangled like the night sky—are so rare that they are threatened both locally and nationally, to varying degrees, with extinction.

This land, sandwiched between two dozen suburban houses, two baseball fields, an asphalt parking lot, and a river, is their only hope for the future. And these women, and a handful of other occasional volunteers, are their only safeguards.

That any turtle egg ever hatches at all is a minor miracle. Everyone eats them—including people. Humans are major predators of turtle eggs the world over. Years ago, the man who carved my wedding ring told my husband that his family used to dig up snapper eggs to make mayonnaise. Traditional Latin and Caribbean cultures celebrate eating sea turtle eggs boiled, tossed raw into glasses of beer, and cooked in omelets. In parts of West Africa, sea turtle eggs are eaten to (unsuccessfully) treat malaria. In Asia, they're (wrongly) considered an aphrodisiac.

It's easy to see why the seven species of the world's sea turtles are all endangered. Only twenty-seven percent of the energy in most sea turtles' eggs ever return to the ocean in the form of living hatchlings, according to the online science library *Faunalytics*; the rest is eaten by predators or broken down into food for plants. The National Oceanic and Atmospheric Administration estimates that the odds of a sea turtle hatchling surviving to adulthood may range from an unlikely one in one thousand to an almost impossible one in ten thousand. Survival rates may be even more disheartening for the less-studied freshwater turtles and tortoises. Some researchers estimate that as many as ninety percent of snapping turtle eggs may be destroyed before the babies are even born.

"I used to find turtle eggshells outside the nests and think, Oh great, baby turtles hatched here," Jeanne told Matt and me. But because hatchlings leave their eggshells behind, in their underground nests, seeing scattered eggshells aboveground means that predators got them. David Carroll told me the sight reminded him of a plundered temple. Skunks, raccoons, foxes, weasels, opossums, coyotes, bears, and crows, among others, all eat turtle eggs—and most of these nest predators are found in higher concentrations in suburbia. So are the curious children and dogs who disturb nesting mothers and dig up their eggs. Even jostling the eggs will kill developing hatchlings.

Far less obvious hazards threaten the eggs as well. Ants swarm nests, awaiting the first cracks to kill the hatchlings. Fly maggots enter the shells and devour the developing babies. Even trees will attack them. In times of drought, thirsty roots invade the nest and penetrate the eggs to suck their moisture. Drought itself can kill the turtle babies inside the eggs. A flood can drown them. Too much heat and they cook.

Even if babies hatch, they instantly face a new gauntlet. Crossing open areas on their way to water and forest, they're snatched by native predators, harassed by dogs, pursued by children, swallowed by snakes, gobbled by crows; even chipmunks will eat them, holding the baby turtle in tiny hands like it's a miniature hamburger and biting off the heads and legs.

"There are so many strikes against turtles," said Emily. And yet the creatures managed to hold their own for eons in a hostile world—until modern humans upset the balance with cars, poaching, pollution, habitat destruction, climate change. "We just wanted to improve the odds."

It all started seventeen years ago. Jeanne attended a talk in her development by a graduate student studying the turtles there. Back before her son was born, she had sometimes seen wood turtles when she walked with her girls, then aged five, three, and one, in the flat, sandy area just past their backyard. Until that night, she didn't realize these turtles were endangered. She had no idea they were nesting. Thinking she was helping, she would carry them back to the river—assistance, she later learned, about as effective as scooping up a woman in labor who's made it halfway across town to the hospital and taking her back to her house.

She wanted to help. "They're quiet and slow and really pretty, gentle creatures. You see one and it's just . . . It makes you feel good to see a turtle," Jeanne said. At the talk, she met a longtime member of the conservation commission, who teamed up with the graduate student and another volunteer to create and deploy the Nest Protectors. Emily joined five years later when she learned of the project from the conservation commissioner. It was perfect timing; as the turtles were nesting, Jeanne was about to give birth to her fourth child.

These wire cylinders, like the one Emily was carrying the day Matt and I first joined the effort, are sunk six inches into the ground, made to fit over the nest, like a chimney, to fence off predators. The protectors worked—until predators learned that the wires marked nests, and started digging under them for an easy feast. The volunteers then added a hardware cloth skirt around each wire cylinder, held in place with large rocks, to prevent animals from digging underneath.

Protecting nests is easier said than done. Emily and Jeanne stay up late nights from the end of May through June and July, waiting for a turtle to lay, to mark the nest for protection in the morning. But they must constantly search for nests they missed. Finding them is difficult. The woodies usually nest first, leaving an almost undetectable swirl in the sandy soil, often on slopes where British soldier lichens grow. Snappers leave a faint trail of parallel tracks and frequently a tail drag—but often dig decoys. "Paints are really hard to find," says Emily. "They cover the nest up so you can never tell they were even there." The eggs are laid pointed ends up, in a clump the size of the O in an OK hand signal, that widens out at the bottom. Blanding's turtles seem to like to nest at two in the morning.

Some years, at the nesting grounds, the ticks are horrendous. One time we were attacked by mosquitoes so huge that Matt and I didn't even recognize them as mosquitoes; we thought they must be some other kind of horrible insect we'd never encountered before.

The busiest times are nesting and hatching seasons, which for the five species here, may span from mid-May through mid-September. But even in between, the Turtle Ladies of Torrington, as Matt and I call them, are busy checking the nests for ants, watering them if too dry, and monitoring for predators.

It's long, hot, sweaty, buggy work, and the Turtle Ladies often cut their hands on the sharp edges of rusty metal. But thanks to their efforts, over the past seventeen years, they have protected many hundreds of nests that would have otherwise been destroyed, enabling thousands of baby turtles to hatch—some of whom are old enough now to dig nests and lay eggs of their own.

So Matt and I were thrilled to get Jeanne's email the night before we were to head back to Turtle Rescue League. The conservation commissioner had discovered the first two snapper nests of the season that Wednesday afternoon, she wrote, and Jeanne had followed four painted turtles around for several hours that evening—one of whom dug a hole and deposited her eggs. "I didn't see any woodies or Blanding's roaming around," Jeanne wrote. "But it won't be too long before they join the party, too. It would be great to get a schedule going. I look forward to seeing you all! Nesting season has officially begun!"

"They'll definitely be out crossing the roads today," Alexxia tells us when Matt and I return to the League the next morning. The nesting wave has hit with breathtaking rapidity.

"It's *happening*!" Matt proclaims. On our drives to the airport, to the aquarium, to TRL, he narrates the landscape in terms of turtles. Even if there are none to be seen at that moment, he will conjure the memories: "See that pond? That's the place I found a bunch of spotted turtles . . . Oh, this is where I always see the first turtles of the year . . . Erin used to take me here when I broke my leg and we would sit and watch the turtles . . . I found a Blanding's turtle crossing the road right here!" All winter, every winter, Matt longs for the sight of wild turtles—the way most people now yearn, during the pandemic shutdown, for a restaurant meal, a cocktail party, a trip to the theater.

"Things are in play right now," agrees Natasha. "Today's going to be a busy day."

In fact, it already is. Scott phoned earlier this morning from the Teamsters Union, reporting more nesting activity. Michaela has been sent off on her first egg retrieval—a turtle is laying eggs in a dirt pile that was dug for an aboveground pool soon to be under construction. To free up hospital space for new arrivals, Alexxia is going to release fifteen baby painted turtles who hatched last year from eggs rescued from patients who came in hit by cars. Natasha, Matt, and I will release fifteen others. Because the League endeavors to return all babies to the same areas their mothers are thought to have come from, that means traveling to five different towns.

Downstairs, as Alexxia counts out babies from the tanks, we greet Fire Chief. He gazes up at us from the water, probably hoping for a banana. (I'd love to give him one, but there's only half a banana in the fridge—and it's too short, say Alexxia and Natasha, for me to safely feed him.) We quickly check on Chunky Chip, who seems to be recovering nicely from the surgery to remove the hooks from his face. We ask about Skidplate: still with us, Natasha tells us. In fact, his wounds might soon be sufficiently healed for him to move from his dry hospital box to a water tank.

And now our baby painted turtles—each between one and a half to two inches long, and very busy—are packed up, sandwiched between clean towels in two twenty-three-by-sixteen-inch white plastic Sterilite bins, securely covered with a lid punched with air holes. Matt, Natasha, and I head outside where my car and the turtle ambulances are parked in the driveway.

"Should we split up, or should we follow you in your car?" I ask Natasha stupidly.

"No, I'm coming with you," she replies.

Right. She's blind.

But using the map in her head and what peripheral vision remains, Natasha will be navigating us toward the release sites for the turtles. Wetlands don't have a GPS address.

"Keep your eyes peeled," says Natasha as we turn on to busy Route 12. Turtle Number 34—a patient who recently died in the hospital—was hit on this road, which is parallel to the river where he probably summered. "You might see a turtle on the road. We can do a little bit of patrol here." Rush hour for turtles, if turtles can be said to rush, starts about five a.m. and slows by ten a.m., she tells us—which, unfortunately, coincides with the human rush hour. And like human traffic, it picks up again at five p.m.

The roads are mercifully less crowded today due to the pandemic shutdown. Months later, a Texas A&M analysis will determine America's commuter traffic was cut by half, and a study by University of California–Davis will show that the numbers of animals, from cougars to turtles, saved by the reduced traffic will be estimated in the tens of millions. Fewer cars on the roads "is the biggest conservation action that we've taken, possibly ever, since the national parks were created," Fraser Schilling, director of the Cal–Davis-based Road Ecology Center, would tell the *Atlantic* magazine. Slowing down America's signature hurry on the highways, compelling folks to stay put for a while, also significantly reduced CO_2 emissions and other pollution—a historic, though temporary, boon for people, animals, and the environment.

But there are ways to reduce road mortality without a pandemic.

Sweden has constructed viaducts for reindeer herds to safely cross highways. Six underpasses along a railway line between Nairobi and Mombassa make way for Kenyan elephants. Banff, in Canada, boasts six overpasses, thirty-eight underpasses, and fencing built to channel wildlife away from its roads—and is estimated to have saved more than 200,000 individual animals, from lynx to toads, since completion in 2017. According to one study, a single two-mile stretch of one Banff overpass reduced crashes with wild-life by ninety percent and saved humans $100,000.

But even small and inexpensive fixes can make a huge difference. Our friend Emily told us that when she used to drive home from teaching school, she often stopped rush-hour traffic to carry a turtle across a state highway near a city line during nesting season. The road bisects a large wetland. She conducted a survey of the site, along a short stretch of one-third of a mile. She counted the carcasses of twenty-nine road-killed turtles in a single day. "That place needed help," she realized. So she and some of her high school students embarked on a yearlong project to fix the problem.

She applied for grants. She amassed fencing materials, the local hardware store giving it to her at cost. When she turned up five hundred dollars short, one of the students' parents wrote a check. Over the course of two chilly weekends that March, one hundred people—teachers, students, and community volunteers—showed up to erect a fence. The owner of a Thai restaurant catered lunch for the volunteers for free. While police cars flashed their lights, volunteers wearing orange vests over their winter hats and gloves buried two-thirds of a mile of knee-high turtle-proof fencing, complete with "doggie doors" for commuting mammals every fifty feet, using zip ties to attach the wire to guardrail posts. If an animal pushed against the door, it would open to the riverside—but not from the river to the road. Though this did limit the creatures' options, the benefits outweighed the risks: The next year, in the six turtle mortality surveys conducted during spring and fall, only one dead turtle was counted.

We find no turtles live or dead along Route 12, but we are headed for an even busier road. Our first release site is just off Interstate 395. We make a few passes to locate the place—it's hard to spot, particularly if you are blind. But incredibly, Natasha finds it. Just at the border between Massachusetts and Connecticut, we pull to the side of the highway and, carrying one box of turtles, emerge from my Prius into the stink of car exhaust and hot asphalt and the sting of the sun's fierce glare off the automobiles and trucks whizzing by. We swing our legs over a metal guardrail, skid down a steep slope . . . and enter a green world of cool shadows, verdant with the fluffy plumes of ostrich ferns, low-growing vines of shiny three-leaved poison ivy, and the upright stems of light green hairy-leaved jewelweed, whose orange blossoms offer sweet nectar to hummingbirds. Yellow-and-black tiger swallowtail butterflies float and flutter through patches of sunshine.

We follow a leafy path along an abandoned 1880s railroad line to the entrance of the wetland. At this moment, a childhood friend of Matt's, appropriately named Random, calls for a FaceTime live report. Random is standing by the Contoocook River in Matt's hometown—right beside a mature snapping turtle. "We're releasing painted turtles right now," Matt tells him. "Good luck!" Random replies.

"Turtles!" Matt cries. It's his hurrah. The very word makes him break into a smile.

"All turtles, all the time!" I reply. And for Matt, that's not far from true. Since his art has become well known, people are constantly contacting him about turtles: sightings of turtles, news of turtles, questions about turtles. Most of the photos on his phone, besides pictures of his and Erin's two dogs, Monte and Roo, are of turtles. Matt confesses that Erin has been known to prep him for social get-togethers by asking him not to steer the talk around to turtles. But he always does.

"What better thing is there to talk about?" asks Natasha.

We wander a bit—Natasha now using her white cane like a walking stick—and soon she finds the perfect spot for the first release. It's along a gentle slope to a beaver pond, bounded by cattails, tussocks, and the long, sword-shaped blades of sweet flag iris, offering the babies ample cover. We will release five of them here, each of us humans squatting five yards apart, reducing the likelihood that a single predator—a frog, a heron, a raccoon, a snapper, a trout—will be able to gulp all the babies at once.

"You're home, kid," Natasha says as she places the first baby in the shallow water. The infant pulls its head in, then shyly pokes it back out, glances from one side to the other, and then, as if possessed by a sudden realization, rushes into the water and swims away.

Natasha's second baby is more confident. After a brief look around, the turtle marches off into the shallows like a shopper at a sale. "It's so interesting to see how each turtle evaluates each situation differently," Natasha says. "Some immediately go on a flavor-tasting binge. Some hide. Some explore." Matt's first release sits like a stone at the edge of the water. Matt waits a minute, then moves it a few feet away. It still won't move. But his other turtle rockets off his palm and swims off like Michael Phelps. Mine is the smallest; it sits with head and legs pulled into the shell, as if overwhelmed by the magnitude of its freedom. I move it a couple of times. But finally, even this shy, tiny baby is ready to take on the world.

At a time when so many of our fellow humans are angry and confused and frightened, how can these baby turtles seem so strong and brave? How can they courageously embrace this world, so full of danger and wonder? I asked David Carroll this once. "That little brain has a few hundred million years of history programmed into it," he told me. "I think they're soil scientists, botanists, hydrologists. They know about all these things. They know what to do. They've lived so long on this earth."

Some might say that baby turtles seem wise because they don't have enough sense to fear their own doom. But David—along with Alexxia, Natasha, Michaela, the Turtle Ladies of Torrington, and Matt and I—looks at this another way: These babies are brave because they are wise, imbued with a wisdom that stretches back hundreds of millennia. That's where wisdom originated—not in some creed that humans made up. For true bravery and wisdom, we have to go back to turtle-knowing, the ancient wellspring that we, too, can draw upon as we struggle to live bravely on the earth.

The babies are now all invisible, immersed in mud and leaves, back in the wild waters from which their mothers emerged last spring. Birdsong rains down on us like a blessing. A male swamp sparrow lets loose his liquid, exclamatory trill; an oven bird calls, *"Teacher! Teacher! Teacher!"* A hermit thrush breathes its sustained whistle and spiraling song. In field guides, the notes are represented in words as if the bird is saying, *"Oh holy holy, oh purity purity eeh, sweetly sweetly."* We turn and walk back.

Abruptly, we emerge from our expedition to the hot car parked along the interstate. It feels like waking from a dream.

But which is the dream, and which is reality?

In modern Western culture, dreams are dismissed merely as stories our brains randomly generate while asleep. In other societies, though, the reverse is the case. In the ancient world, dreams brought messages from the gods. Mesopotamian kings paid special attention to dreams. Dream divination was a common feature of ancient Greek and Roman religions. For ancient peoples from Amazonian hunter-gatherers to the authors of the Holy Bible, dreams carried profound meaning. In this special state of altered consciousness, ordinary people could break free of the bonds of the present and time-travel: Dreams foretold the future.

The Aboriginal people of Australia evoke what is called, in English, "Dreamtime" in their art, culture, and religious celebrations. It's a concept difficult for outsiders to understand. "Dreamtime is a beginning that never ended," explains a website for an Aboriginal art gallery, Artlandish, in Kununurra, Western Australia. "Dreaming" is the word that explains the ongoing creation and actions of the animals, plants, landscape, and spirits of the earth. "Dreamtime is a period on a continuum of past, present, and future," the website continues. The word was coined, of course, by a Westerner. A white man based in Alice Springs came up with it, the best translation he could muster, and it was popularized in the 1890s by the British-born anthropologist Walter Baldwin Spencer. But none of the hundreds of Aboriginal languages historically contain a word for "time."

In Aboriginal cosmology, Dreamtime is a realm that transcends linear time. In Christianity, that realm is known as eternity. Hindus may call it Moksha; Buddhists, Nirvana. The Greeks coined the word *kairos* for sacred time, and envisioned it as an eternal spiral through which we escape chronological time. Many physicists and philosophers say these religions have the right idea. They say separation between past and future is fiction; in their view, everything that was and is and is to be "is contained in a vast university of eternity that swells but does not move," as Lewis Lapham writes in the foreword to an edition of *Lapham's Quarterly* devoted to the subject of time. Einstein agreed: He believed, as the astronomer Michelle Thaller, assistant director of science communication at the Goddard Space Flight Center explains, "that the Big Bang created all of space and all of time at once, so every point in the past and every point in the future are just as real as the point in time you feel yourself in right now." In other words: "You, right now, have been dead for trillions of years . . . you haven't been born yet . . . everything that's happened to you, if you could get the right perspective on the universe, you could see it all at once." Einstein, Thaller says, postulated that time is like a landscape. Past, present, and future are all spread out before you, whole. "The distinction between past, present, and future is only an illusion, even if a stubborn one," he insisted. And there are situations, continues Thaller, in which time essentially does not exist. "At the speed of light," she explains, "time does not progress at all. Light does not experience time."

Yet *time* is the most commonly used noun in the English language. Time must be counted, like money, and like money, it runs out. According to our clocks and calendars, time is an arrow—shooting inevitably toward our doom. In my sixties, I am increasingly aware of this, as new wrinkles, and new aches in joints, begin to appear. It is even more evident in the many friends older than my husband and I, folks who seemed ageless but who now walk more stiffly, stand more slowly, and forget things. I don't fear my own death (I will join the dogs and ferrets and birds and pig and turtles and octopuses who went there before me, whether into peaceful nothingness or Heaven or another dimension)—but I do fear, mightily, the deaths of those I love. Yet there is nothing I can do to avert this. Time goes in one direction, moving, in the Greek philosopher Heraclitus's famous image, like a flowing river into whose waters no one can step twice. Or like an arrow, the weapon maliciously lodged in the neck of innocent Robin Hood.

And yet, time also heals and restores. At dawn, it's not the clock radio my husband sets to the news on NPR that wakes me these days; it's the song of a wren out our bedroom window that eases me out of my dreams, proclaiming a nesting territory, as wrens do spring after timeless spring.

Two kinds of time exist side by side: the frenzied, fleeting, harried time, rushing along like the cars on the interstate, and the eternal, cyclical, renewing time of the seasons. The turtles traverse them both. Following them to the world just outside the highway guardrails, we enter the embrace of the wild, beating heart of nature, escaping for a moment from the trap of transience.

We can't be exactly sure where the mothers of all our hatchlings came from. Freshwater turtles travel farther than one might expect. Though no match for a record-holding leatherback sea turtle whose telemetry recorded a journey of 12,774 miles from Indonesia to Oregon in 2008, female snappers have been logged traveling as much as ten miles away from their pond; a 1983 radio telemetry study of paints in southern Saskatchewan recorded one turtle traveling four miles in a single day. How do they know where to go?

To find their way on long-distance ocean journeys, sea turtles use a magnetic compass. Two, in fact. University of North Carolina researchers discovered that, from birth, these turtles possess one magnetic sense that detects the angle at which the geomagnetic field lines meet the surface of the earth, and another one that detects differences in the field's intensity. These two senses deliver information similar to a sailor's longitude and latitude, since most spots in the ocean have a distinct combination of the two.

It's not known whether all turtles navigate in this way, or whether they use other senses. But we are careful to return babies to the town where they were found, because location is important. Also, we want to keep intact the natural gene pool. But when possible, we hedge our bets by releasing them in several locations and slightly different habitats in the same wetland.

Natasha steers us into a housing development of McMansions. We park along a dead-end road. We can't help but stop right next to a driveway—there's nowhere else to leave the car—where a man is getting out of his vehicle. I wonder if we three strangers, carrying our large mysterious box, should introduce ourselves to the homeowner and reveal our mission. But he doesn't even give us a glance.

All of the newish homes turn their backs to a cedar swamp, as if pretending it isn't there. "It's a secret, hidden world," says Natasha. "No one's even looking at the place where all the life happens."

Assisted by Mr. Stickey, Natasha leads us past the houses and to the edge of the cedar swamp. It is a wetland cathedral, canopied with conifers, scented like pencil shavings, and carpeted with sponge-like sphagnum moss that cushions and squelches with every step. Soon we are up to our calves in the acidic tea-colored water, as floating islands of moss sink beneath our feet. An iridescent skim of duckweed and heart-shaped lily pads dot the dark surface. Here we release five more babies.

And last, we drive to the town of Auburn. Its 16.4 square miles host no fewer than four interstates, including Routes 90 and 290, which form an X at its center. Somehow—how does she *do* this?— Natasha directs us to a suburban neighborhood ablaze with planted purple rhododendrons and yellow bearded irises. She finds the perfect spot to release the last five hatchlings, with plenty of basking rocks, downed logs, a gentle slope to the water. Natasha's hatchling rests for a moment in her hand, head out, breathing in the scent of the warm air, black eyes sparkling. "Welcome home, kids," Natasha says to the brood. And then she reminds Matt and me: "If it weren't for our efforts, these little guys would have been just rotting eggs."

But instead, from the moment they step from our hands, they begin lives that connect them to a tradition that stretches back millennia. For these turtles, at least, the world has, for the moment, been set right again.

Natasha checks her messages. They are piling up. There are several reports of injuries, some from other states. Nearby, a potentially gravid painted turtle is in rough shape, having been hit on Route 49—a highway that cuts through wetland after wetland. A volunteer will be bringing that turtle in. By then Michaela will be back from rescuing the eggs to receive her. Meanwhile, we head to the Teamsters Union. Scott has reported one active nest already, and by the time we arrive, there may be more.

It's three thirty when we turn up the drive to the parking lot, and we're greeted with orange cones, posted with several Xeroxed signs announcing, TURTLE SEASON IS HERE! and PLEASE USE CAUTION. "Scott really takes this seriously," says Natasha, "and I love him for that." The signs are illustrated with a picture of a box turtle—the one whose domed shell seems decorated with hulled walnut halves, the one with the hinged plastron that closes up tight, the one that everyone recognizes. This species doesn't nest here. But snappers and painted turtles do.

We drive to the back of the big lot and walk over to see where Scott has marked a recent nest with a piece of hardware cloth. It's only six inches from the asphalt lip of the parking lot. But just two yards away, a painted turtle is poised, her head largely cloaked in her fleshy "turtleneck." What she is attending to is behind, not ahead. Teetering at first toward the right, then the left, slowly she wags from side to side. "She's doing the nesting dance," whispers Matt.

We approach quietly. We don't want her to abandon her task. Sometimes, Emily told us, mother turtles seem to go into a nesting trance. This trance is well documented in sea turtles. In his 1979 book *Time of the Turtle*, Jack Rudloe notes, "Once turtles begin to lay, they are oblivious to all activity around them. Even people blasting off electronic flashes in their faces or shining lights on them doesn't make any difference. Not even thumping their shells causes them to stop."

What is this like for the turtle? One interesting guess was imagined by Stephanie Ondrack, a Canadian birth doula, after witnessing a giant green sea turtle lay her eggs in the hole she had laboriously dug on a Costa Rican beach. Birth doulas support human mothers in labor, often at home. "Like the turtles," she writes, "human mothers settle deeper and deeper into a trance-like state as the birthing hormones slowly infiltrate the brain." These hormones, she writes, "soften the edges of her perception, blur her thoughts, and infuse her with coping abilities she would not otherwise possess." If the hormones are allowed to reach their natural heights, she says, the laboring mother will break free of the passage of time, lose awareness of the others in the room. She won't be able to answer questions. She will want dim lights and quiet. Her trance will allow her to focus deeply inward, to concentrate entirely on the process of giving birth.

What is the experience of a turtle in a nesting trance? We cannot know, because turtles cannot tell us. But human mothers can. And the many women who Ondrack helped as they gave birth at home were able to tell her later about their experience. She writes that they described an ecstasy as if "you could blend the feelings of having just won the marathon, climbed the mountain, received the Nobel prize, had sex with your true love, and experienced a religious miracle, all rolled into one sensation of success, empowerment, elation, passion, the sacred, and blinding love." This is not an intellectual exercise. It is entirely mediated by hormones—the same hormones present in nesting turtles.

That most women in labor never experience this elation, Ondrack insists, is because the birth process and its hormonal cascade are repeatedly interrupted in today's medicalized maternity wards. Similarly, a nesting turtle who is interrupted before she begins to lay can be deflected entirely from her task. So we give this turtle her privacy. "Laying might take several hours," Natasha warns us. "We can't do anything till she's done."

Remaining at a respectful distance, we scan the weedy, sandy slope past the mother, looking for more activity. From the highway, trucks roar past, a distant ambulance shrieks, a car radio blares rap music from open windows.

"Wait," whispers Matt. "I think I hear a turtle walking."

We peer through the unmowed grass and tall weeds of the slope. Several yards in front of a chain-link fence that demarcates the Teamsters property from the adjacent conservation land, we spot a low, rounded form, about five inches long. Is that the turtle? Keeping out of the line of sight of the nesting mother, I tiptoe closer, slowly, heart pounding, for a better look.

No, it's a rock.

Meanwhile, the painted turtle continues her nesting dance: a laborious, hypnotic rocking. "Can you tell if she's digging or tamping?" Natasha asks.

"Tamping," answers Matt with certainty. He's seen turtles nesting many dozens of times. Digging is a more energetic affair. Often you see dirt actually flying. It may be, some scientists have theorized, that just as humans in various cultures have induced trance states through physical exercise, ordeal, or dance, in turtles the physical labor of digging the nest helps induce the nesting trance. But the movements we are watching now are far subtler and more gentle, as first one strong back foot, then another, presses the earth back into place over her precious eggs. To further compact the soil, Natasha explains, she wets the nest with liquid from her bladder—which is why, she warns, if you help a female cross a road, you must handle her with great gentleness. A turtle who feels threatened often releases urine to startle or deter predators, but a nesting mother who has to do this on her way to the nest must then detour for another drink of water before she can resume her task.

The mother turtle starts to walk away. With the exception of a small handful of species, mother turtles, like most mother snakes, allow the earth to incubate their eggs, and do not return to the nest to check on the developing embryos. Laying the eggs is labor enough. This mother must be exhausted, and in this state more vulnerable to predators. "I'll give her a ride back to the river," Matt volunteers. Barefoot again, he gently lifts her up and carries her down to the water's edge. On the way there, Matt spots the other turtle—which he did indeed hear walking earlier. She's another painted turtle, who is now resting in the sweet, cool shade of fern and pine.

It's painfully obvious that both nests are too close to the parking lot for safety. To position myself downslope of the nest Scott marked, I must sit on the asphalt, where I can feel my skin sizzle through my lightweight pants. On a 77-degree (F) day, the temperature of asphalt can reach 125. At 85 degrees, blacktop can exceed 140. This would be terrible for the eggs. In most species, temperature determines the sex of the hatchlings, with the cooler temps producing males, and the higher ones turning hatchlings female. But after a certain point, you get boiled eggs. One study conducted at the famous nesting site at Costa Rica's Ostinal Beach found that when the air temperature soared above 95, of the tens of thousands of eggs laid there by olive ridley sea turtles, not a single one hatched.

While Matt excavates the paint's eggs, Natasha directs me as I dig up the lower nest. Matt has done this many times. Paints typically dig nests less than four inches deep. His agile artist's fingers know by muscle memory the contours of the painted turtle's flask-shaped cavity. He recognizes instantly the cool, smooth oval curve of the typically four or five eggshells, and starts removing them to the padded egg transporter with careful efficiency.

But I am nervous. Mine is a snapper nest, Natasha tells me, and to reach the topmost eggs, I may have to dig half a foot or more, past rocks, roots, dirt, and sand. I am fearful of piercing an egg. This is most likely to happen at the very moment you finally locate the nest cavity, as your finger pops through a hard layer of compacted sand, dirt, and rocks to the chamber itself. "It's not just a hole," Natasha tells me. "It's a structure, with definite walls, and even a roof. You'll know immediately when you find it." I have dug down five inches and now am brushing away less than a quarter inch of sand at a time, using only the tips of my fingers.

And then the first glimmer of shell appears. Until this moment, I didn't fully believe they would be there. The eggs are perfectly round, about the size of Ping-Pong balls, and just as white. I look at them in startled wonder. I feel like a person who has never before looked into the night sky, who suddenly beholds—hanging *right there!*—the full moon.

I gently lift the first egg I find, careful to keep the sphere perfectly level as I transfer it to the transporter, lest I slosh the contents and kill someone who could live over a century. And next to it, the sand reveals another, and then another . . .

The mythologies of widely disparate times and cultures hold the egg in reverence. Each egg is a new beginning, its smooth shell enclosing a self-contained universe, its roundness reminding us of the circle of life. To this day, honoring a tradition that predates the religion, Christians give Easter eggs to celebrate the resurrection of Jesus. (It seems appropriate: Even though He did not hatch from an egg, we are told that He did rise, like a hatchling turtle, from a hole in the ground.)

The mother snapper had packed her eggs firmly and carefully in the ground, stacked neatly as cordwood in rows. Matt has finished moving the paint eggs: They number seven. But I am still digging. "Care to guess how many?" Natasha asks me. I think a moment. "Ten?" I guess. She laughs. "Keep going."

I'm grateful my hands aren't shaking with the gravitas of the moment. Sanskrit scriptures tell that all existence began with an egg. Egyptian mythology, in one version, says that the sun god, Ra, ruler of order, of the underworld, of kings, and of sky, hatched from an egg—while half a world away, in Australia, the Aboriginal people explain that in Dreamtime, the sun was created from an egg hurled into the sky. Greek orphic tradition holds that the primordial deity, Phanes, hatched from the original egg, itself created from Fate and Time.

Even today, cosmologists look to the egg to show us how science explains the beginning of the universe. Nearly fourteen billion years ago, the theory goes, the entire mass of the universe was compressed into a "space-time singularity" of infinite density—which, upon its explosion, birthed all that we would come to know: space, planets, suns, matter, turtles, people, and time. But long before that theory came to be called the Big Bang (the term was originally derisive), a Belgian priest named Georges Lemaître, in a 1927 paper in *Nature*, attempted to explain the concept of an expanding universe. He called that original, infinitely dense singularity the Cosmic Egg.

I've uncovered twenty eggs by now, and I'm still digging. At the edge of the parking lot, my seat and thighs are baking. Sweat drips off my nose. Ants crawl up my hands, onto my arms, and under my shirt. "The ants are another reason we need to take these eggs," Natasha reminds me. Her voice seems strangely far away. It is as if I've entered a reverse nesting trance. Nothing in this world could be more momentous, more fulfilling, or more joyous than excavating these eggs.

Finally, I lift what is clearly the last, precious egg from the nest. There were thirty-one.

"Great job, team!" says Natasha. With one bin filled with eggs, another emptied of our charges, we drive back to the hospital feeling victorious.

As we turn into the League's drive, a blue Prius pulls up in back of mine. A young man gets out, holding a six-inch-long, dark-shelled turtle in both hands. The guy lives in the neighborhood, and while driving past, he had taken note of the bright-green house with the Turtle Rescue League cars parked in front. He looks hugely relieved to have found us.

A glance at the turtle's orange-red plastron and dark rounded carapace tells us he's got a painted turtle, and the short front claws at the end of her yellow-striped arms reveal she's female. She doesn't look injured. Why bring her here?

"I saw it crossing the road," he explains as his charge air-swims in his hands, "but there was nowhere I could take it. There's no water, no grass, just paving . . ."

He doesn't know—and who can blame him?—of the oases existing just outside our own perceptions. Right in back of suburban backyards, sometimes inches from asphalt parking lots, these most ancient of vertebrate animals are reenacting the miracles that keep the world alive.

Natasha gratefully takes the animal from the kind man's hands. "Don't worry," she assures him. "We'll find a good place for her." Later that afternoon, she and Alexxia walk from their door to the wetland in back to set her free.

Painted turtles bask in the sun's warmth.

7·

Fast and Slow

Chunky Chip reclaims his pond

Bob Garfield was crying, and it broke my heart.

With his coanchor, Brooke Gladstone, Bob Garfield hosted the weekly National Public Radio show *On the Media*, to which my husband and I, trained as journalists, often listened each Sunday. "There's this thing that's happened, at least to me, as the pandemic plays havoc with our present," Garfield told his listeners that spring. "Not knowing what the passage of time will yield has left me unmoored, spinning in space as if my inner gyroscope were on the fritz.

"Time isn't just a metric," he said. "It's a gravity that keeps us tethered to the world." Disconnected from time, he felt lost. "Losing the sense of future, I have also lost the present. . . . Truthfully," he confessed, "I've been crying. A lot."

I'd always thought of Garfield as a tough guy—at least a tough-minded reporter—so the thought of him weeping distressed me doubly. But what was worse was that so many of our fellow citizens were suffering that same despair. Later, on a different Sunday, the lead headline on the cover of the *New York Times'* Sunday Styles section announced "Everything's a Blur" and promised to tell "how isolation, monotony, and chronic stress are serving to destroy our sense of time." It went on to describe "the paradox of 2020, or one of them: a year so momentous also seems, in a way, as if nothing happened at all." Psychologists explained to the author, Alex Williams, that there were many reasons for this. Facing the uncertainties of a mysterious killer virus, political chaos, environmental catastrophe, and racial unrest creates a state of chronic stress so severe, it interferes with the brain's ability to form memories—a brain fog not unlike that which sufferers of Long Covid experience. Memories, of course, are the way we order our fundamental experience of time and change, the way we anchor our sense of self in the flow of life. But what if nothing seems to change, from day to day, week to week, month to month? What ensues, then, is "a collapse of the reassuring feeling [that] our lives move in orderly progression" so essential to human mental stability. We become like a turtle stuck on her back, helpless, unable to move.

Young people on the cusp of setting out miss milestones of proms and graduations; few could land jobs or internships in the shutdown; many have to live at home with their parents, a rerun of childhood. Michaela, at eighteen, had graduated from high school less than a year when, as she put it, "time froze. I wasn't in high school. I wasn't in college. I was just at home with my grandma and my cousin."

For elders stuck in nursing homes, it's even worse. Denied visits from family and friends, one dreary day blends into another and another, like the hopeless soliloquy of Macbeth upon learning his queen is dead: "And tomorrow, and tomorrow, and tomorrow, / Creeps in this petty pace from day to day, / To the last syllable of recorded time; / And all our yesterdays have lighted fools / The way to dusty death." For so many people waiting out the crisis, time has lost its boundaries, and life is drained of meaning.

But when we are with the turtles, our experience of time—in fact, our experience of almost everything—is completely different from those of our fellow countrymen. Michaela's girlfriend, Andi, for instance, feels caught in the pandemic time warp. She had hoped to find some direction taking photography in college, but Zoom classes were lame, and now she doesn't know what to do with her life. But for Michaela, working with the turtles gives her "calm, stability, and a sense of purpose": "I've dived into something that's really meaningful, doing something to help a living creature."

Thanks to the turtles, we are profoundly immersed in spring's unfolding, and deeply connected to the progression of the dramas in the turtles' daily lives.

Natasha's strong arms are straining to hold Chunky Chip in position. "He's probably the third largest snapper we've ever had," she says. He weighs forty-eight pounds, so big and strong that Alexxia and Natasha need to use an electric screwdriver to secure, and then to remove, the screws holding the lid over his gigantic stock tank shut.

Natasha struggles to maintain the Wheel Well Grip on Chunky's shell while she sits on the swivel stool near the operating table. His armored, fifteen-inch-long tail hangs down between her legs. Then, from an opening on the underside of his tail, Chunky Chip extrudes a huge, purple, seven-inch-long tubiform appendage tipped in what looks like a club. It resembles a fantastic, elongated mushroom you might expect to find growing on Mars.

It's his seven-inch penis. "A sight to behold, and one that can startle both novice and experienced herpetoculturalists alike," writes M. Honda as quoted in a blog post titled "The Terrifying Sex Organs of Turtles" in *Scientific American*'s *Tetrapod Zoology*. (And yes, all turtle penises are shockingly huge—sometimes half the length of the carapace. "The evolution of the shell probably means that male turtles were forced to evolve innovative penises in order to make genital contact with their partners," muses Darren Naish, the paleozoologist who authored the article. But sometimes, for unknown reasons, a turtle drops his penis in response to being handled as well.)

Alexxia, though, is interested in Chunky's opposite, and much sharper, end. While Natasha holds the snapper up, Alexxia uses a gynecological tool in her right hand to pry apart his jaws so that with her left she can insert a champagne cork. Because of turtles' slow metabolism, anesthesia is avoided when possible—both because it takes so long to take effect and because recovery is more difficult. So the cork is needed both to keep the mouth open and to prevent the turtle from biting.

Then Alexxia's cell phone rings. She holds it between her shoulder and ear. "Hi, Mom!" she says. I can hear her mother's cheerful chatter on the other end. Mothers are used to hearing their grown daughters say they're too busy to talk, but Alexxia actually has a good excuse: She's doing oral surgery on a very large, wild, and fully alert snapping turtle.

Yesterday, Alexxia drained an abscess from the puncture wound left by the most recent of the fishhooks she's removed from Chunky's lower jaw; there are now pockets of pus inside his mouth, too. She touches one with a dental instrument. Chunky lunges, nearly dislodging the cork. "I've got a turtle right now, Mom," she says into the phone. "Let me call you later, OK?"

All week, the phones at Turtle Rescue League have been ringing off the hook. "Last night we passed out, exhausted," Natasha murmured through her mask when we arrived that morning in early June. Though we've only been gone a few days, so much has happened.

New patients this year, including those who have died, now number fifty-three. The dog-chewed painted turtle, named Tacos, is not doing well. The back of her shell smells like dirty feet, which indicates her wounds may be infected. But incredibly, Skidplate, the big snapper who'd been hit and dragged beneath the car, continues to recover. Despite extensive injuries to his plastron, tail, and cloaca, his digestive system is working again, as a recent poop full of pine needles from something he'd eaten before his injury attests. Another huge snapper, dubbed Scratches, who came in last week, laid seventy-two eggs last night. She pushed out the last twelve as Natasha and Alexxia watched, at nine p.m. Then at eleven thirty p.m., the couple got a call from someone who hit a snapper in the road and was too afraid to touch the animal they'd injured. "We gave the number for a local rehabber," Natasha told us, "but we heard nothing from the rehabber we'd recommended." The driver may never have called, leaving the wounded turtle either on the road—where it would be hit again and again—or to wander off and, unless found and rescued, probably die. We may be beating the bushes for that turtle today, Natasha told us. If there's time.

The first order of business, though—while Alexxia is at work fixing machines and Michaela is on the way from Rhode Island, bringing eggs from a snapper who nested in a pile of loam dropped off for someone's flower bed—is to pack up three nine-month-old snapping turtles for release.

To start their new life, we head to a cemetery.

The memorial park abuts conservation land that had been generously given to the town by the same donor. It's perfect habitat for our overwintered babies. With Natasha leading the way with her white cane, we plunge from the manicured lawn into blackberry thickets, descend past discarded plastic grave flowers and fifty-milliliter liquor bottles, and into a hemlock forest that grew up after farmers' fields were abandoned in the late nineteenth century. Our footsteps are cushioned by princess pine and hay-scented fern, and the moist, morning air swells with the spicy scent released by the ferns' crushed fronds. We follow a stone wall to a large, shallow pond, where the residents of a big beaver lodge have provided plenty of sunning logs for turtles: To our delight, Matt immediately spots the shiny black shell of a basking paint. The shy reptile dives. "He feels our eyes on him," says Matt.

To the trills of tree frogs and the burp of a bullfrog, we line up, five feet apart, and each release a hatchling. Mine, the largest, nearly four inches long, rushes into the water, swims five strokes, and hides beneath the mud. Matt's rests under a tussock, then burrows under. And Natasha's "is just one of those philosophers, soaking it all up," she says. She looks away for a moment—and the next thing she knows, he's gone.

We walk back through the young forest, elated. But before we even leave its shade, Natasha's cell phone rings.

"Turtle Rescue League. How can I help you?"

It's a skateboarder who's encountered a mother snapper digging a nest alongside a bike path in Rhode Island. He's worried about the mother. He's worried about the eggs. What should he do?

"Are the eggs in the hole, or are they scattered?" Natasha asks, holding the phone in one hand and wielding the white cane in the other to navigate over a fallen log.

"The best we can do is rake the loose dirt, cover the eggs, and pat it with the palm of your hands," she says. The guy's happy to do so. "Thank you for helping out this turtle!" Natasha says warmly.

We get in the car and shut the doors. But before we have a chance to leave the cemetery grounds, the phone rings again.

"Turtle Rescue League. How can I help you?"

"Yep, yep, sure," she responds. Skateboard guy has called back. Now the turtle is actually on the bike path. He's afraid she'll get hit. "Are you aware of where your closest water is?" asks Natasha. That's where the mother's likely headed. "Find a thick, blunt branch. If you're afraid of being bit, nudge her off the bike path. After having given birth, she needs a place to rest and catch her breath." She explains the Platter Lift if he's up for trying it. But he's not.

"Good, good, perfect!" Natasha encourages patiently. He's using his skateboard to encourage the turtle to move. "She honestly just wants to get home at this stage," Natasha explains. "You're genuinely a turtle hero today! Thank you so much!"

Next, as we drive back toward HQ, Natasha checks the League's Facebook page. Multiple calls and messages are piled up. Reports of turtles in need come from other states, other nations, other continents. Every year they get calls about how to help road-injured gopher tortoises, a southern species from southwest Louisiana across to Florida. A message has arrived from Albania; a couple of tourists have found an injured turtle but no vet will treat it. They don't know what species it is. People call in reporting injured turtles, sick turtles, nesting turtles—and turtles who are just freaking people out. One man in Belmont, a town just west of Boston, calls urgently to state that when he opened up his garage door, to his dismay, a turtle came "rushing" inside. He's too afraid of it to move it.

12:40: "Turtle Rescue League!" It's the skateboard dude again. "Yeah, you really don't have to walk the turtle all the way back to the water," Natasha reassures the turtle's protector. "It's good she's off the trail so people won't bother her. You can now wish her well and just let her complete her journey. Sign off—case closed!"

12:52: "Turtle Rescue League! How can we help you today? A turtle egg situation. We'll send someone out. Can you give us your loc—"

The signal dies.

But they call back. "So what's the situation?" A turtle of unknown species has nested, as often happens, in the dirt delivered for a construction project. "Did the turtle lay just today? Text us your address. We'll send someone out. It might take a few hours." In fact, it will: The address is in Lanesboro, far western Massachusetts, two hours' drive away. Natasha texts Michaela. She's on it. But first she has to attend to a report of a turtle hit by the side of the road in Brookfield.

2:19: "Turtle Rescue League. How can we help you today?" It's the animal control officer from the town of Wilmington. A turtle has been hit by a car, and eggs are spewed out on the road. "OK. We have a volunteer in Newburyport, only twenty minutes away." That's Mike Henry, who brought in Robin Hood. Natasha dispatches him to check the carcass for internal eggs. He'll retrieve any salvageable eggs for his own incubators. That's a good thing. TRL has five incubators. Each is a modified 120-quart Coleman picnic cooler holding plastic food storage containers with trays of eggs on top of sterile soil. The containers' lids are propped at an angle with clothespins, creating a sloped roof. Warmed by an adjustable fish tank heater and humidified by an inch of water at the bottom, the incubators maintain a temperature between 80 and 86 degrees, and the five of them may hold as many as a thousand eggs. But all five are already nearly full.

2:30: A text from Michaela, which Natasha directs her phone to read aloud. It does so in double time, because Natasha can process hyperfast speech. The machine-driven message sounds like it's being spoken in monotone by an insectoid robot. It even translates emojis. "It's-a-very-dead-turtle-I-nearly-puked-by-the-side-of-the-road-grimacing-face."

2:39: A longtime volunteer, Dianne Dougherty, calls about a turtle hit by a car in nearby Dudley. At least one egg was seen on the tarmac.

2:46: Back at TRL, Dianne's red Ford pulls up. "She's coming apart," she mumbles beneath her mask. We peer inside the corrugated cardboard box where the patient rests on a pink towel. It's a paint. Her carapace is shattered. Her bridge, the piece of shell connecting carapace and plastron, is crushed. Red blood and yellow egg yolk soak the towel. Yet her head is out, and her eyes are open.

Natasha dictates a text to Alexxia at work: "Dianne's turtle is here. Really tough condition." We take the turtle downstairs to transfer her to a clean towel and a hospital box.

Male wood turtle.

Left to right: *Natasha, Matt, Alexxia,* (in front) *Sy, and Michaela.*

Cris Hagen holding an alligator snapping turtle at his house.

A repair to the broken plastron on a painted turtle at Turtle Rescue League.

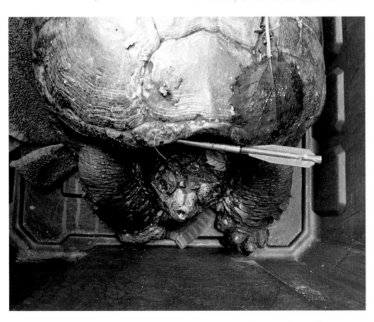

Robin Hood, a snapping turtle who was shot with an arrow.

Clint Doak holding a slider at Turtle Survival Alliance.

Matt demonstrates the Wheel Well Grip on Fire Chief.

Left to right:
*Matt, Emily, Erin,
and Heidi check a
nest for eggs at the
nesting grounds.*

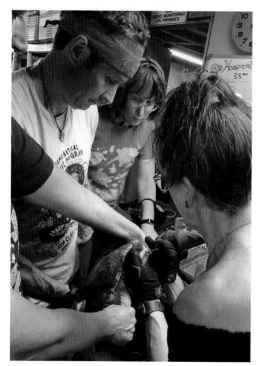

Left to right:
*Matt, Natasha, and
Alexxia work on a
patient at Turtle
Rescue League's
exam table.*

Matt and Sy kayaking at Turtle Cove.

A wood turtle from the river behind the nesting grounds.

A nest protector at the nesting grounds thwarts predators.

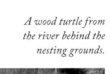

A baby painted turtle hatched from Turtle Rescue League's incubators.

Sy unearthing snapper eggs at the edge of the parking lot at the Teamsters Union.

Chunky Chip, aka Tortzilla, about to be released at his home pond after recovering from his fishhook injuries and infection.

Scott holds an infant snapper before it's released back to the wetlands at the Teamsters Union.

A baby spotted turtle.

During a break in his physical therapy session in the Turtle Garden, the gentlemanly Fire Chief accepts a cuddle from Sy.

Heading out on the Cape for the nighttime sea turtle rescue. Within hours, the sled will be filled with five endangered Kemp's ridley sea turtles and the heavy seaweed that will insulate them on the journey.

The Chief attempts to breach the wooden turtle excluder wall by the front door to greet his friend Sy.

A winter afternoon lounging in sun puddles with Sprockets, Pizza Man, and Apricot in Alexxia and Natasha's living room.

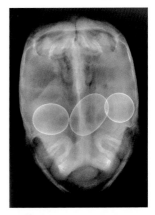

*Lucy's X-ray reveals
an egg is stuck.*

*Erin immediately
picked Addie, a
friendly three-toed
box turtle, as the
newest family
member of the
Patterson household.*

*Sy overjoyed that
Lucy is eating again.*

Fire Chief enjoys a cuddle from Matt and Sy.

Lucy's treatment restored her to radiant health.

A parade of children and parents file toward the pond to release their headstarted hatchlings.

Natasha and Alexxia after releasing Nibbles, their first turtle patient at Turtle Rescue League.

Releasing a baby Blanding's into the river behind the nesting grounds.

Matt seizes the giant Asian river turtle for weighing at Turtle Survival Alliance.

Matt with his teenage African spur-thigh tortoise, Eddie.

Matt and Sy up to their necks in turtles at Turtle Survival Alliance's Survival Center.

Matt digging a turtle pond in his backyard.

The new turtle pond at Matt and Erin's house.

Swimming with the Chief.

Nesting eastern painted turtle.

2:58: Alexxia arrives in record time from a service call. She looks like an anime action figure, clad in sleek black Lycra and military-style high boots, her cell phone strapped to a boot garter like a gun in a holster. Alexxia lifts the lid on Dianne's turtle. "Hello, little monkey . . .

"Oh. Ouch. Bad." She lifts the turtle onto the exam table and flicks on the light. She pulls the neck out with her left hand and holding a dental tool in her right, pries open the mouth, checking for blood clots.

"All right, little lady," she says. She holds the turtle at an angle and squirts Ringer's solution along the crushed bridge to clean the wound. She uses another dental tool to push pink flesh that has gushed through the broken bridge back into the body. She pulls the two sides of the turtle back together and secures them with super-glue and aluminum foil tape.

"Will this turtle survive?" I ask timidly.

"Honestly, it's an extreme long shot," she replies. "I'm just trying to get her where, if miracles are gonna happen, they can happen. But everybody gets a chance here."

3:15: Injected with painkillers, antibiotics, and fluids, the crushed paint rests on fresh towels in a hospital box now labeled in marker on masking tape with her weight, 370 grams, and the number 54. Alexxia turns her attention to the turtles in bins stacked around the exam table five-high awaiting their follow-up checks and meds. But first she takes Spunky, a large snapper with a head injury, out of the water bin she shares with Scratches. She sets Spunky on the cement by our feet. I flash back to my years as a medical reporter: I often attended surgeries, but never was there a large reptile running around loose on the operating room floor.

"Chill out," she says to the turtle, and pats her on the top of her head, which tilts markedly to the right. The big snapper takes off toward the stairs with the air of an absent-minded professor on a half-forgotten errand.

It's good for Spunky to walk, Alexxia explains, to let the turtle get back her mental bearings. "If you engage them and stimulate them, it helps them recover better," she tells me. She interrupts Spunky's journey to gently squeeze her feet, then her tail. When she steps away, the turtle resumes her task—and Alexxia resumes hers.

4:10: Everyone in the tower of bins is carefully checked. Number 36, a painted turtle, gets her shot of ceftazidime and a drink of water from Alexxia's cupped hand. Number 52, a snapper who came in two days ago with a fracture on the back half of the shell, receives a shot of fluids. Number 47, a paint, has her bandage changed, which looks like a fresh diaper on a baby. Number 53, a very large old painted turtle with a break in her bridge, is persuaded to drink on her own, from Alexxia's hand. She won't need a shot to hydrate her. Number 44, a snapper with a traumatic head injury, receives some antibiotics . . .

4:30: Alexxia slides out another bin, much larger, from underneath the exam table. It's Number 27: Skidplate. Natasha Wheel-Wells Skidplate up onto her lap so Alexxia can examine his scraped plastron. When he was dragged along the asphalt beneath the car that hit him, his plastron, tail, and back legs became what motorcycle riders like Alexxia call "a road crayon"—leaving a long smear of flesh along the road like a trail of pigmented wax on paper.

"He is doing so great!" Alexxia pronounces. She cheerfully cleans some poop off his large tail. "I like that his poop is coming out of the right place," she says. "I was really worried about that tail wound." Snappers need their tail, as well as their head and neck, to right themselves if they're flipped upside down, but of more immediate concern is possible damage to the digestive tract. Unlike in mammals, where the gut ends beneath the tail, in turtles, the opening is on the underside of the tail itself, as are the sex organs. "Good for you, Skidplate! Cases like these," Alexxia tells us, "you really get attached to."

"He's already stolen everyone's heart," agrees Natasha.

. . .

All summer long, turtles keep stealing our hearts. The personable red-footed tortoise, Pizza Man, and his Burmese buddy, Sprockets, continually surprise and delight us. They both recognize Matt and me by now, and seem to enjoy it when we stroke their outstretched heads. Downstairs, we always check in with Percy, the centenarian box turtle, who has lost none of his spunk to age. When we clean his habitat and change his water, we lift him out to run after us on the floor. There's a spotted turtle we call Mr. Pajamas because he lost perhaps a third of his yellow-spangled shell to a car accident; the shell is growing back beautifully where it attaches to the flesh beneath, but not beyond—and most of it is still pinkish. With his dark legs sticking out immodestly without the usual overhang of his carapace, he looks like he's not quite dressed.

For their distinctive appearances, for outstanding personalities, and for all they have endured, we are taken with so many of the turtles here. The brain-injured turtles, like Snowball, Scratches, and Spunky, hold a special place in my heart. All the big red-ears are packed with personality. We love the diamondback terrapin, Corndog, who has grown so fat she's practically busting out of her shell. Matt particularly loves Apricot, the elongated tortoise who hangs out with Michaela's favorite, Pepperoni. "She has such a kind face," he says. He loves to pick her up and hold her, and when she looks at him with her dark eyes, it's clear she recognizes him and enjoys his attentions.

But of all of them, Matt's and my favorite is Fire Chief.

We always look in on this huge, ancient snapper in his giant tank on its high shelf. For me, it's a little disconcerting; Matt has no trouble peering in because he's six feet tall, and both Alexxia and Natasha are tall as well. But I'm only five foot five, and with Fire Chief's tank up so high, to see him at all I have to put my face inches from his—a position from which he could, as we saw when he grabbed a banana, lurch from the water like a croc. But he never does.

"In a whole room full of turtles, he's special," Matt observes. That is saying a lot. What makes Fire Chief so distinctive?

Fire Chief gazes back up at us. His eyes track us when we move. He is interested. Though his shell was badly smashed and his legs and tail were paralyzed in the accident, he has no head injury. His brain is perfectly fine. He is an absolutely magnificent, huge, ancient snapping turtle, and he knows it. Although he's been in the hospital two years, he is still utterly wild. What he has, Matt says, is "a wild appeal."

And yet, Fire Chief deigns to favor us with his attention. Matt and I are eager, when there is time, once the nesting season is over in early July, to start helping him with physical therapy. He might— just might—regain full use of his back legs and tail. Alexxia feels his chances of a complete recovery are remote. But she and Natasha agree: We owe him, and all the eligible turtles here at TRL, the chance to return to their natural homes, and to thrive, or to struggle, or to live for another century, in the wild.

One morning, we arrive at the rescue anguished, with a female paint we'd picked up at Emily's house near the nesting grounds. The dog belonging to a neighbor across town had used the turtle as a Kong toy: From a dog's point of view, the hard exterior of a turtle provides pleasant chewing exercise, and if they work on it long enough, a secondary reward of a delicious, meaty treat in the middle.

The little female's carapace is in ruins, huge sections of both front and back shell chewed down to the flesh. Her plastron is broken in several places and hanging on by shreds. But she's still very active. As Alexxia works on her, the feisty paint struggles, trying to get away. She bites Alexxia's thumb, drawing blood. "Ahhh!" Alexxia yells. "That hurt!"

"She wants so much to live," Matt whispers.

"I love the fighters," says Michaela.

But cases like these, Alexxia tells us, seldom make it. "They don't survive because the injuries are so extensive, and there's introduction of bacteria from the dog's mouth."

Still, Alexxia tries. She injects some lidocaine, a topical painkiller. She irrigates the wounds. She tries to put back what is left of the chewed and crushed shell. She administers fluids. She injects a large dose of antibiotics. Dr. Barbara Bonner, the veterinarian who had owned Percy, had success in last-ditch cases by keeping the turtle cool for the first few hours of treatment. Alexxia is going to try that in this case.

"So that's Turtle Number Seventy," Alexxia says, closing the top of the hospital box. "Let's see if she wakes up tomorrow. That's as good as it gets with dog chews."

Natasha, as always, tries to cheer us up. "The good news," she tells us, "is that Skidplate is still doing well!" It's been a month since his injury, and he has passed an important milestone: Alexxia and Natasha have found the big hurdles to survival typically occur at three days, three weeks, and again at three months. Skidplate has already passed two of these critical landmarks.

"I literally am so connected with that kid," Natasha confides, "that I have to hold his front foot when we give him a shot of Baytril. His pain reaction is like an uncontrolled muscle tremor. I feel it. I tell him, 'Kid, we're just doing the best we can.' And I think he can feel our connection, and it helps."

Soon, Skidplate won't need those painful shots anymore. He's been on antibiotics for four weeks, and his wounds are closing nicely. That means he'll graduate from a box lined with towels to a watery habitat more like home. "He's just a week away from water time!" says Alexxia.

Sliding the big box forward from beneath the exam table, Alexxia opens the lid eagerly.

"Hi, honeypie!"

But Skidplate doesn't move.

His eyes are sunken. The Doppler shows no heartbeat.

Natasha strokes his feet. "Oh, kid . . ."

"He must have passed not long ago," says Alexxia. "*That* really sucks. He was supposed to be our miracle kid, too."

We all fall silent. I want to cry, but I don't. If tears are to be shed, they rightfully belong to Natasha and Alexxia and Michaela—not me and Matt, who are dilettantes in comparison. But we, too, needed him to live. We needed Skidplate to help us buffer the heartbreak of the many, like Number 70, who we will lose. At a time of plague and violence, of climate-killing pollution, of human greed and overpopulation and problems for which solutions were willfully ignored, we needed this turtle, at least, to triumph over the odds.

"The world is at least / fifty percent terrible," the poet Maggie Smith wrote in "Good Bones." "For every bird there is a stone thrown at a bird. / For every loved child, a child broken, bagged, / sunk in a lake. Life is short and the world / is at least half terrible . . ."

But after the poet's litany of sorrows comes her defiant refrain: "I keep this from my children," she writes. And why? She answers this in the next line: "I am trying / to sell them the world."

She is hell-bent on loving this gorgeous, terrible, heartbreaking world—so in love that she loosed upon it creatures made from her own body, delivered in great pain: offspring who will, eventually, repeatedly, break her heart (for isn't that what children do when they grow up and leave?). Yet she births, and loves, and, as best she can, shields her children—all so that they, too, might have a chance to witness, along with the sorrow, the miracles.

Every year, Alexxia promised us, there would be at least one "miracle kid." Just a week after Skidplate died, we celebrated one of them.

Matt, Alexxia, Natasha, and I teetered across a stone wall and finally perched on a log overhanging a protected pond full of blooming white water lilies—the perfect spot to watch the turtle explore the water where we set him free. It was Chutney—the "roller" with the head injury so hopeless that a veterinarian had suggested euthanasia as the only humane option.

For four months, he literally didn't know which end was up, and every time he flipped, Alexxia had to reset his broken jaw.

Today, this hopeless case is healed and free. Chutney walks without even the slightest head tilt. In the water of the pond, his soft, wrinkled neck stretches out smooth, surging forward with such eager intent that it appears almost a separate creature. He spots, then seizes, then swallows a bug—his first wild food in more than three years. It must taste delicious.

July 7, 2020: The number of Covid-19 cases in our country has reached three million, with eight hundred deaths a day—a new record. Again. This is the fifth such record in nine days. The next day, though, we turn our attention away from the pandemic and immerse ourselves in turtle time.

Snapping turtles in various stages of recovery are crawling all over the basement floor. I lift Spunky out of the way so Matt can heft Chunky Chip out of his tank and move him to a fifty-gallon transport box. Matt's knuckles run with blood where Chunky's two-inch claws rake him during the lift. Chunky may not even know he's done this; turtles have a tendency to air-swim when they are lifted, as if they are reflexively hurrying away from a predator.

The transport box completely fills the trunk of my car. Even though we secured the locking lid with an extra bungee cord, as we pull out, I see the box rocking ominously, and we hear the lid clicking as if it is about to pop open. Matt's driving now, with Natasha in the passenger seat, so I am the first line of defense should a huge beaked head, followed by the strong armored front feet, suddenly erupt into the cabin. "It's been known to happen," notes Natasha—who, never being the driver, is always the one who has to capture the errant snapper and stuff him or her back into the box at the earliest possible moment. Though once, she remembers, an escaped snapper on the way to release spent the car journey riding on top of his transport box. The turtle watched the passing scene with interest out the back window as other drivers and passengers cheerfully waved.

Mercifully, by the time we head onto the Mass Pike, Chunky has calmed down. But we haven't. We're excited: Fully healed, Chunky's returning to the pond he has called home for a century. But we're also anxious, because we're sending him back to the very place where he's been seriously and possibly intentionally injured three times in two years. Can't law enforcement help? Natasha tells us, to our dismay, that while most other native turtle species are protected by law, capturing and killing snapping turtles—by trap, hook, spear, gun, or even arrow—is perfectly legal in Massachusetts year-round. "I hate the thought that you can care for them for so long," says Matt, "bring them back to health, and somebody can just come along and kill them."

Happily, Chunky's friends outnumber his enemies. In Marblehead, where he was known as Tortzilla, he has a big fan club. The neighbors will be waiting to welcome him home. The story of his injury and recuperation made the local papers. One family has phoned TRL regularly to check up on him. Last week when they called, they were worried about another big shell they could see just breaking the pond's surface. On closer inspection, it looked like two turtles, one on top of the other. They were concerned that another male snapper had not only taken over Tortzilla's territory but was moving in on his girlfriend. Natasha reassured them: Most ponds have several snappers, and many DNA studies have shown that a single snapper clutch can have as many as five different fathers.

As we pull in to our destination, we're greeted by Nancy and Phil. Theirs is one of the several stately homes lining the landscaped edge of Tortzilla's two-acre pond. "He'd hang out here," Nancy says, blue eyes crinkling above her mask, "and every day he'd pop his head up and stare into my eyes." The retired couple will be moving soon, downsizing to a house in town. But first they want to see their old friend—and make his home as safe as possible. People don't fish from their end of the pond, says Phil. But at the opposite end, there's a clearing where he plans to pull a big branch into the shallows, to make casting a line difficult. They often hear kids playing down there, adds Nancy, and Phil plans to give them an education about turtle stewardship.

Soon two other neighbors appear. Bridgette, wearing an elegant dress and speaking in a French accent, and her husband, Paul, are also friends of the big snapper. They live in town, but know Tortzilla through Nancy and Phil, so they drove out for the occasion. Shortly another fan, Peter, will be arriving, and taking photos and video of the event.

Matt and I heave the weighty travel box into the backyard, past plantings of hosta and hydrangea cascading down along stone steps that lead to the water's edge. While we wait for Peter, we learn more of Tortzilla's story.

When Phil and Nancy first moved in thirteen years ago, occasionally they'd see the big snapper at the dock. He'd sometimes climb up on the rocks, as if to investigate the people. "At first I thought he wanted to bite us," Nancy confides, "and I was sort of afraid of him." But the reptile's persistent curiosity won them both over. "He knows us," insists Nancy. "And I've fallen in love with him!" They started feeding him bananas five years ago, which he eagerly accepted—but Nancy noticed that even when they had no fruit to offer, Tortzilla would still appear. "It's not just about the bananas," she insists. "He loves me." And it's a family affair: A big female they named "Mrs. Tortzilla" laid eggs one year in the dirt along their stone driveway. When twenty-seven babies hatched out, Nancy carried each down to the pond for release.

"He's a remarkable individual," Natasha agrees. "He is very curious. Even with all the procedures we've done, few of which were

very pleasant, he was always eager to see who was there to see him."
Last summer, Nancy and Phil noticed the first hook in Tortzilla's
jaw. With the help of a younger couple that was visiting, they managed to haul the huge turtle out of the water and into a box, and
drove him to TRL for treatment. The hook was big: not some kid's,
but a poacher's gear. It took three weeks for Tortzilla to recover. He
was released during a downpour, and Natasha and Michaela celebrated with Phil, Nancy, and other guests with cheese and crackers
and cold drinks. Just two weeks later, Nancy called the rescue again.
A bobber attracted her attention to a smaller hook stuck in Tortzilla's neck. But before they could recapture the turtle, the hook was
clearly gone. What they didn't realize was the hook, bobber, and
attached line had been torn out, creating a slit across the turtle's
throat that over the summer, fall, and winter would become pro-
gressively and dangerously infected. Nancy and Phil noted Tortzilla
had emerged from hibernation a few days before Memorial Day.
A few days later, they noticed a new fishhook in his mouth. "If
we hadn't rescued him," Nancy asks Natasha, "would he not have
survived?"

The small fishhook, Natasha explains, might have just rotted
out of Tortzilla's mouth. But the infection from the previous hook,
untreated, might well have killed him.

Turtles, though, excel at battling infection on their own. When
they feel sick, they take action to effect a cure. Turtles fighting illness may spend more time out of the water, courting the heat of the
sun to raise their body temperature to fight the germs. Sometimes
sick turtles even emerge from their winter hibernation to bask in
the sun's healing apricity.

"How can we make sure he doesn't get injured again?" asks Paul.

"Well, I hope he'll be more wary of people," answers Natasha.
"We're hoping cutting back on feeding him snacks teaches him that."

"We'll speak to the new owners and make sure they're not too
friendly to him," Phil promises.

Peter, a younger guy in a Palm Springs T-shirt, has now arrived with his camera to record the release. Matt hoists Tortzilla up out of his box and places him on the grass at the edge of the water.

"Oh, Chunky," pleads Natasha, "please be careful!"

The huge turtle runs into the pond. Almost instantly he's swallowed by the cool dark water, and he disappears.

"Just think of him going from that tank in the hospital to this," whispers Matt.

The eight of us stand at the water's edge, silent. "I see him!" shouts Phil. "That's him, where the trail of bubbles are," Matt says, pointing. "Welcome home!" cries Bridgette. "I'm so glad he's back," says Nancy. We smile at each other from beneath our masks.

Just then, an older lady with a white West Highland terrier comes upon the scene. Nancy walks to meet her, and though we can't hear what they say, it's clear the exchange was unfriendly. Nancy returns to us and says tersely, "She hates snapping turtles." The woman and her husband, she explains, believe snappers eat baby ducks and geese.

"But he's mainly a scavenger!" says Natasha indignantly. "He can get much of his diet just from the bottom of the pond. He's not going to take geese. Hunting takes a lot of effort." Snapping turtles do enjoy protein, and youngsters eat a lot of bugs, larvae, and small fish. But adults mainly feast on carcasses—in the process, helping keep ponds like this one clean.

Nancy relates that the terrier owner's husband would probably like to shoot Tortzilla. Of course, discharging a gun in a residential area, I note, would probably get him arrested, and wouldn't be worth the effort. But her crankiness has us worried.

"When we released him last year," Natasha later confides to Matt and me in the car, "it was one hundred percent victory. I went away from that release feeling great. It was devastating just two weeks later to have him hooked again. When we heard the hook pulled through, I thought, Oh, just a glancing blow . . .

"I certainly shared the joy of watching him charging into the water. But I'm not going to say the world doesn't look like it's becoming a scarier place. The pond seems a lot smaller. But at releases, you're always plagued with uncertainty."

Now, at the pond's edge, though, Natasha keeps her fears to herself. The trail of bubbles shows Tortzilla has reached the middle of his pond. "I hope his wits and skills are enough to protect him," she says.

And then, as if in reply, an osprey wings across the pond. Sometimes called a fish hawk, this raptor is bigger than a goose, with narrow wings and long legs adapted to seizing fish from the water. "I consider it a very good omen from Mother Nature," says Natasha. "At age one hundred, Chunky Chip—or Tortzilla—may only be middle-aged. I want to think about this turtle, long after we're gone, still ruling his beautiful pond."

8.

First Steps

A painted turtle digs a nest

On a shady ridge above the river, Emily, Jeanne, Matt, Erin, and I are sitting on springy pillows of pine needles, looking down on Torrington's turtle paradise.

Two stories below, logs felled to construct a beaver family's two dams offer safe platforms on which an adult painted turtle and a sizable wood turtle are basking in the hot summer sun. On this perfect blue-sky day, the air itself seems to shimmer with life. Ebony damselflies, little dragonflies who hold their wings back like Tinker Bell, flutter past sandy banks where greenbriers unfurl slender, spiny tendrils. Chipping sparrows surround us with their long, dry trills. A great blue heron oars the sky with six-foot wings.

Suddenly, the basking paint dives into the cool water. With binoculars, we see she's chasing a school of minnows. Now we spot two other swimming turtles—wood turtles, Emily realizes at a glance. "One's coming right toward us!" says Matt. Not far away, yet another turtle is intently exploring the sandy bottom of the tea-colored river. "That's a BIG snapper!" says Jeanne with admiration.

"What an awesome spot," says Erin. "I want to be a turtle."

It feels good to rest here after a morning of helping Jeanne and Emily water the nests. This has been the hottest July on record in the entire Northern Hemisphere, and one of the driest ever in this corner of New England. This has meant that to prevent the eggs from cooking or drying out, every three days somebody has had to trudge up steep slopes, often through sand, and under a merciless sun, lugging thirteen one-gallon jugs of water, each weighing more than eight pounds, to the forty-six protected nests now scattered over the Torrington nesting grounds' twenty-five acres.

This nesting season has been particularly hectic. After a slow start, one June day, after a rain, suddenly "turtles were everywhere," recalls Emily. "Woodies, snappers, paints, and even a Blanding's!"

"It was like Jurassic Park," Jeanne's husband, Bryan, said. One night, "There were thirty turtles out there—and the great blue herons flying overhead like pterodactyls." Jeanne was out for ten full hours that night, waiting for turtles to finish digging so she could erect protectors for their eggs. ("My whole body ached," she said. "I felt like telling them, 'GO HOME!'") Emily was there, too—and for many other nights like that one. Sometimes she wondered if the turtle she was watching had turned into a rock. She watched one Blanding's dig a nest for forty-five minutes—she went on to dig four more. As it turned out, she laid eggs in none of them.

We'd seen turtles fool us before. One Sunday morning in June, I'd brought my twelve-year-old friend Heidi Bell along to help us. A ten-pound snapper, sand on her head, was digging her nest on one of the slopes. We withdrew, checking for other nests, to give her time to finish. Half an hour later, we returned to find her resting, looking exhausted, a few steps away from the hole she had carefully filled in.

"Good job, honey!" Jeanne said to the turtle. "This nest will be the eleventh snapper nest of the year so far," she said. But so as not to waste a nest protector, and to properly place it if deployed, the Turtle Ladies always check for eggs first.

The mother turtle seemed unconcerned about the seven humans who surrounded the nest she had so carefully constructed and filled. While Heidi looked on in wonder—she had never been so close to an adult snapper before—Emily, Matt, and I started to dig at 9:56 a.m. Gingerly, careful not to injure an egg, we used only our fingers. "Who's going to find buried treasure first?" Emily asked encouragingly. We dug down three inches, four inches, five. After six inches, all seven of us got involved.

By 10:15, the hole was three feet long and one and a half feet deep. "This hole is big enough to hold me in it!" Heidi said. Still we had found no eggs. The mother turtle remained sitting three feet away, as inscrutable as a sphinx. "Is she just messing with us?" asked Matt. "This is craziness. I'm starting to doubt . . ."

"Maybe it was false labor?" Heidi wondered. (Her mother is a nurse.)

"This is insane," said Bryan. "Are you ready to call it?"

At 10:20, Emily and Jeanne officially pronounced the nest a ruse. "We were conned!" I said to the motionless snapper. Matt plucked a large leech from her left rear foot, and still she didn't move.

The next morning, Jeanne found a new snapper nest full of eggs—just a few yards from the hole we had spent nearly half an hour excavating. We're pretty sure it was the work of the wise mother snapper, whose convincing decoy had successfully fooled the most lethal predator species that has ever lived.

"Turtles always manage to surprise us," Emily says as a blessed breeze ruffles the surface of the water below, cooling our skin. "That's part of the joy of it." In the seventeen years they've been protecting the nests, Emily, Jeanne, and the other volunteers who help have witnessed many wonders: Turtles who they thought too tiny to breed laying eggs. Turtles who made nests in slopes so steep they had to lay practically standing on their hind legs. Turtles born too small to survive—who made it, nonetheless.

Turtles have led the volunteers to other revelations as well. One evening when Emily was checking the slopes for nests, she spied a little semicircle of stones and broken bits of green maple leaves near one of the baseball fields. "It clearly hadn't gotten there by itself," she told us, so she lifted the leaves to find, beneath them, a newly hatched baby bird, naked and shivering and cold. She picked it up, not wanting the infant to die alone, to take it home. As she was leaving, she heard one of the boys she'd noticed hanging around the baseball field cry, "It's gone!"

Emily kept the baby alive that night on a heating pad, feeding it crushed blueberries with an eyedropper. In the morning, she called a wildlife rehabilitator. Emily didn't know what species the bird was. A robin? The rehabber didn't have a permit to treat robins. But maybe it was a sparrow? The rehabber had a permit for these, and took the orphan in.

Four days later, the baby opened its eyes. It was chirping. It had grown feathers. They were blue—it was a bluebird, its color symbolic of peace and contentment, a creature widely loved as a harbinger of happiness, a promise of renewal each spring. The rehabber named the bird Glee, and the bird grew strong enough to be released back to the wild.

Emily wrote a note she left on the ground, anchored by a rock, where Glee had been found. It was addressed "To the boys who rescued me":

"I'm doing fine," it said. "I'm a bluebird. Thank you for rescuing me."

With the busy nesting season over, a sense of calm settles over the Turtle Rescue League's electric-green house. Pizza Man is still sleeping in his shelter in the yellow-tiled bathroom when Matt and

I arrive at ten. Sprockets has recovered from a fit of pique, after Natasha took him out of his soaking tub too early for his liking. (Offering tortoises a good soak keeps them well hydrated and prevents bladder stones; also, they clearly enjoy it.) After his untimely eviction, Sprockets stomped around, angry, for hours, Natasha tells us, knocking things over and wedging himself into corners. But now he's back to his easygoing self. His reward will be to play a game of egg baseball with an unfertilized boiled duck egg from a neighbor's farm, still in the shell. When he bites at it, his horny beak slips over the smooth surface and sends the egg spinning away. But because it is ovoid, the egg shoots along the kitchen's pink-and-blue tiled floor and then comes looping mysteriously back. Sprockets's eyes shine with delight. He bites, steps, strikes—sending the egg wobbling off on another erratic orbit. And on it goes, for five minutes and twenty-eight seconds. On his sixth try, Sprockets finally has the egg cornered. It cannot escape. He crunches through the shell, his beak slicing into the rubbery white and the custardy yellow yolk. His satisfaction is so big it engulfs us all: At once, we four break out with a satisfied "Mmmmmmm!"

But the focus of our efforts on this late-July day will be Fire Chief. Today we begin his physical therapy. Matt and I have been growing ever fonder of him, always stopping at his tank to exchange glances as he pulls his giant head from the water to find our faces.

Matt hefts the huge turtle over the lip of the tank and into a travel case on the cement floor. Fire Chief's bulk completely fills the bin. To carry him safely up the stairs, we try to snap a lid on top—but his enormous head pops the seal and pokes out, looking like the T. rex from the movie *Lost World*. With both Natasha and I pushing hard to urge his head back down and hold the lid shut, Matt walks up the stairs from the basement, through the living room, and out to the deck. Here he lifts the huge, scrabbling reptile out of his container and hauls him over the three-foot wooden fence to the Turtle Garden.

Fire Chief is even more magnificent out of water than in. His head is as big as Tortzilla's. His neck is muscular, not fat. His fourteen-inch-long tail sports eleven tall, proud reddish-brown osteoderms—toothlike bony ridges rising skyward, rather like the spikes on Stegosaurus's tail, though his are not sharp, but rounded. Some of them are a full inch high. His shell is an unusual, gorgeous reddish brown, the color Matt knows from his artist's palette (and I from the 1966 deluxe Crayola pack) as burnt sienna.

His shell tells the story of the catastrophe that brought him here. A Good Samaritan saw the truck run him over. "It was no accident," says Natasha. He flipped on his back, then rolled down an embankment and fell into the fire pond where he usually spent his summers, several hours' drive from the League's HQ. By the time Natasha and Alexxia arrived with their kayak, the entire fire department had turned out—concerned about, but also intimidated by, their snapper friend. The men watched in awe as 130-pound Alexxia dove in and wrestled the injured giant into the little craft.

Fire Chief's shell still looks carunculated toward the front third, where a huge avulsion has sloppily healed over like crashed-together tectonic plates. But the worst damage is beneath the shell, farther back and unseen. "There is no doubt that the fracture pulverized the vertebrae and spinal column," Alexxia had told us. For months, his back legs trailed uselessly behind him.

Natasha scratches the back of his shell and Fire Chief's rear end wags comically in response, but it's no laughing matter. It's a reflex response, Natasha explains, like a shiver, and it's telling: In response to her touch, the nerves from the spine are firing, but they may be doing so randomly, like a pinball crashing around, rather than relaying a message in orderly fashion. That Fire Chief can move his legs at all is a minor miracle. And it's one that gives her hope. But to recover, he will need months, and possibly years, of supervised physical therapy.

"The idea today," Natasha tells us, "is to let him explore, for him to experience full gravity." He seems eager to do this. The Turtle Garden is about the size of a big elementary school classroom, featuring grass, vinca, ferns, and flowers, fallen leaves, a mulberry tree, a blueberry bush, tunnels lined with sand to crawl through, and, atop a small, grassy hill called Mount Olympus, a shallow waterfall and a small pool hopping with frogs. Fire Chief's neck stretches out fully, and his powerful, scaly forefeet pull him forward five, ten, fifteen, twenty steps, heaving him toward the edge of the fence . . .

His mind is obviously clear. No head injury here. He is curious, active, focused. But the back of his plastron is dragging, whereas a snapper without injuries walks high on his legs, belly shield off the ground. The front of Fire Chief's plastron is held high, but the back is dragging.

But importantly, his back legs *are* moving. As his powerful front legs and claws grip the ground and pull, his injured back legs alternate to help push his heavy body forward. They just aren't strong enough to lift up the back end of his body.

After twenty-five steps, Fire Chief stops to rest. Two minutes later, he's on the move again, avidly exploring along the wooden fence. His neck is stretched out so far, it looks like he can't wait to get ahead. Most snappers don't move like this, but keep their heads partially pulled into their turtlenecks.

"This is his first outing of the year," says Natasha. She's carefully assessing his every movement. Though she is officially blind, her retinas still are sensitive to movement, and she knows well the gestalt of a healthy snapper's stride versus that of a turtle with a neurological injury. "He's clearly not paralyzed, but his spine has limited signal. We need to get his rear end up. I have no doubt his muscles have atrophied." Of course they have; ever since his flesh and shell wounds healed, he's been in water, rendering him weightless.

"I would love to get him back into the wild," she says. "You can't have a sufficiently large tank for a snapper. His world should be acres. With a long-term injury, he's been seeing the four walls of his tank way too much. We've been at a plateau with him. So unless we redesign his therapy, he won't get better."

Fire Chief paces steadily along the fence line, rounds a corner, and tackles a five-degree uphill slope. "Keep up the good work, kid!" calls Natasha, now settled into a patio chair. But he's no kid, and Natasha knows it. He is a wise elder: majestic, finished, and complete. He is my age, or even older—and suddenly, surrounded by people so much younger than I am, I feel very proud that I have this in common with Fire Chief: We are both old.

Age is not celebrated in our culture. My friend Liz, who lived among the San Bushmen in Namibia in the 1950s, reasons that in a time and place where our kind need no longer fear attacks from hunting lions or tigers or pumas, aging itself is the ultimate predator. For this reason, we seek to conceal it, dyeing gray hair, Botoxing wrinkles, surgically stretching our sagging jowls and chins—lest death stalk and find us.

For the San, it was different. "Old" was an honorific: In their language, the world for "old," *n!a*, is also the term used when speaking of gods, a word used to convey respect. One has earned a prize when one has attained old age in a culture such as theirs; rather than seeing life ebbing away, the San see in their elders life that has accrued. Like elephants and orcas and who knows how many other species, they realize that the old have a treasure chest of stories and knowledge that the young, for all their energy and enthusiasm, simply cannot be expected to have amassed. Matt remembers watching a video that claimed human vitality peaks in the thirties. Immediately, he knew that was wrong. Maybe that's true for some kinds of physical prowess, he reasoned, but it's not true of the life of the soul: "I know my art is better than it was before. And I'll be better in ten years. And you, too," he told me reassuringly. "We're kind of like turtles. The more we experience, the better we are!" Vitality does not decline with age for an artist. Nor for a writer. And certainly not for a turtle.

"Old things are better than new things," Kami Garcia asserts in her gothic novel *Beautiful Creatures*, "because old things have stories in them." Fire Chief certainly has stories. And we are helping him write a new chapter.

"I'm sure it's frustrating for him," says Natasha. "He's a big male turtle in his prime, and has lived strong and free for sixty years, maybe a hundred. But for some reason, the world is more difficult now."

Matt and I are sticking close to the giant old turtle, watching his huge scaly feet and inch-long nails grip the ground. We're scanning the substrate ahead for sharp objects that might cut his plastron. We're making sure he doesn't flip over—the damage to his legs and tail make it more difficult to right himself. So far he has been walking on soft, fresh grass and composting brown leaves. He's chugging along, impressively fast for a turtle. But he's still dragging his back end. There's a rock sticking up out of the ground ahead. "I'll help you!" I tell him—and a wave of tenderness and gratitude shoots me back in time.

Once again, I am helping my seventy-six-year-old father, stricken with lung cancer and weakened by chemo, as he tries to climb the stairs to the bedroom. He was an army general who as a young officer had survived the Bataan Death March and years as a prisoner of war of the Japanese. As a child I had ridden on his shoulders; I learned ballroom dance steps with my feet on his shoes.

Once again, I am lifting up our elderly border collie Tess when she would wake in the night, confused and unable to rise, to set her right so she can get her bearings before I carry her downstairs to go outside. A rescue who had overcome abandonment and a crippling accident before we ever met, Tess was a hero, too, who taught me to play Frisbee, who always waited for me on trails so I would not get lost when my long-legged husband would hike ahead, who had given me comfort and joy unbounded for fourteen of her sixteen years.

"I'll help you!" was the only weapon I could hurl at their ailments. But it was a strong one, and a privilege I was endlessly grateful to wield.

To be able to help a cherished elder incites a different kind of satisfaction than, say, consoling a crying baby or picking up a fallen child. To help a new life, in all its potential, is an invocation; to aid an old one, in its fulfillment, is a benediction. It is a supreme honor to be able to give back some measure of comfort to those who had nurtured and inspired me. Perhaps the craftspeople who practice the Japanese art of kintsugi, or "golden repair," feel the same way. Kintsugi is an ancient technique of repairing broken ceramics. Rather than trying to make the vessel seem new, rather than attempting to conceal the breaks, the craftsman bonds the broken edges together with a sealant dusted with gold, silver, or platinum. It reflects the philosophy of wabi-sabi, embracing the idea of aging and imperfection, celebrating the beauty of broken things, honoring the effects of time, and giving the gift to the artisan of being able to make a repair. It reminds me of the care and love with which Alexxia repairs turtles' broken shells; it reminds me of the scars on Fire Chief's back.

Gently, respectfully, I lift up the rear of Fire Chief's shell so he doesn't graze the rock. He keeps going a few paces past it, but then stops. The work of walking on land is exhausting. Matt and I so admire him.

After another two-minute rest, Fire Chief turns around and begins striding downhill, neck way out, eyes sparkling, drinking in the familiarity, and the novelty, of this outdoor world thirstily through all his wide-open senses.

"He's certainly determined," says Natasha. Fire Chief has now arrived at a different area of the Turtle Garden, hugging the fence, and approaching more rocks. "In a few minutes, we'll do a plastron check," Natasha announces.

For now, he heads up the gentlest slope of Mount Olympus and plunges into the shallow pool. He quickly crawls out and back down the slope. Matt and I hover like helicopter parents. He heads for the fence again and I help him over another rock. We stop for the plastron check. Though like the carapace, the plastron is a bony structure (nine bones in all), it can be injured by scrapes—which is why turtles don't drag their plastrons on the ground as they walk if they can avoid it.

Matt holds Fire Chief up, the plastron facing toward Natasha and me for inspection. To our relief, we see it's smooth and unscathed. Fire Chief is surprisingly patient. He doesn't gape; he doesn't lunge; he doesn't snap. Matt puts him back down on the ground and the turtle stands still.

And then Matt and I spontaneously feel the same urge at once. Something compels us to do what nobody with any sense should attempt: reach out and touch the front end of a giant wild snapping turtle we barely know.

I am not thinking about him murdering a banana. I am not remembering watching Robin Hood lunge at Alexxia when he had finally had enough. My mind does not flick back to the turtle who was hissing and snapping while I tried to get her across Route 101 on my drive back from the airport. My mind is just filled with this particular individual at this particular moment, and how much Matt and I care for and admire him. Fire Chief has enjoyed his time outside with us as much as we have. We have made a connection.

Matt and I each reach out a hand. Matt gently strokes Fire Chief's neck, and I touch the surprisingly soft skin near his armpit. "You're a big banana cream pie," Matt whispers. Finally, with our fingers, we stroke his mighty head.

Matt and I gave this no thought; it was purely instinctive. Why do we reach out to touch the faces of those we care for? The face, in all animals who have one, is full of touch receptors. At least twenty different types of nerve endings have been identified in skin, and they involve far more than sensing just heat, cold, pain, pressure, and vibration. Among the newest discoveries is that humans have special nerve endings called c-tactile fibers, which respond only to gentle touch. But these kinds of nerve endings had been discovered in other animals in 1939. They are now understood to be "caress detectors." That we and our fellow animals have evolved nerves that hunger for gentle touch is a powerful testament to how important it is to all of us. Across taxa, scientists have found that gentle stroking activates the natural opioid system, the body's chemical pleasure palace.

The bright expression in Fire Chief's eyes changes to a dreamy look, as if he is diverting attention from what he sees and wants to dwell instead on the tactile. He obviously enjoys the interaction. We primates, addicted to the visual, only rarely concentrate on our other senses. But we close our eyes sometimes when we are savoring a delicious taste. We close our eyes while we kiss, and also when we pray. Cognitive scientists tell us this frees our brains to focus on the other, nonvisual aspects of the experience. I think we also do it because it makes us more vulnerable—and thus more open to our most ancient shared languages, to touch and to trust.

The three of us give ourselves to the moment. For Matt and me, it brings rare relief: to enter a world without clock, without calendar, without speech, without worry—lost, and found, in the pleasurable communion of two species, skin on skin, in turtle time.

Together, we rest. And then, restored, the big turtle gathers his strength and decides to continue his walk along the fence. Now we see that something wonderful has happened: He is carrying his plastron entirely off the ground! He turns, and smashing over a hosta next to one of the blueberry bushes, heads to investigate one of the tunnels. He doesn't enter—and this is fortunate, because I'm not sure we could get him out—but turns again to cross the center of the garden and rests in the shade.

"I'm thinking we might establish a regular therapy day," muses Natasha. We could include other turtles in the rehab sessions, too.

Snowball appears to have turned a corner, says Natasha. She and fellow snapper Special have become friends, and now often sit together in a shared bin with one's hand resting on the other's or both hiding together under the floating carpet that stands in for lily pads they'd find in the wild. Special might enjoy a day out as well.

After nearly a year without touching the ground, Fire Chief has been out walking for forty-five minutes. "I think he's wiped," says Matt. It's time he returns to his tank. It's my turn now to pick him up. To my delight, he doesn't thrash or struggle when I lift him. Perhaps because his back legs are exhausted, I can easily hold him with both hands on either side of the middle of his shell and carry him horizontally, which I instinctively feel he would prefer to the Wheel Well Grip. Rather than put him in the box, I just carry him down the stairs, the back of his shell pressed against my belly at an angle, so he can still see ahead. Matt lifts him into his high tank. "We'll be back soon," I tell him. "We're going to get you well," Matt promises. Though he can hear us, we know Fire Chief can't understand our words. But some kind of understanding has clearly transpired between us. He looks up, directly at our faces, before we replace the screen over his tank. We head home, full of hope and promise.

"Every day, the first thing to do is check the incubators," Michaela tells us. "What a way to start the day: seeing tiny, adorable baby turtles!"

We're eager to get Fire Chief out again—but if anything is worthy of delaying an appointment with our big, ancient friend, it's watching new turtles emerge from their eggs.

By the middle of August, eighty-five baby turtles, both snappers and paints, have already hatched in TRL incubators—including those from the eggs Matt and I dug up at the Teamsters Union. They await release in nursery bins filled with sterile mulch—some burrowed under, others marching about, crawling enthusiastically, some piling in corners on top of others like a turtle pyramid.

As Matt and I look on, Michaela opens Incubator 4 to check its six trays, all from central Massachusetts. "Look for pips and slices," she instructs, "and for deflated and 'drooling' eggs."

A pip is the first sign an egg is hatching—the first tiny hole, created by the baby turtle's "egg tooth," a temporary sharp projection on the beak, which is later resorbed. The turtle next enlarges the hole, and may create a slice with their tiny needle claws. A deflating egg means a turtle is coming out of it—often with slimy birth juices that look like drool.

Immediately, Matt spots one round snapper egg with a hole— and lifting it, spies a tiny eye staring back out at him. "Welcome, little one—we're glad you're here!" I say, wondering what he thinks of his first glimpse of this world.

Hatching can be a fraught affair. Once a baby erupted so prematurely from TRL's incubator that he was translucent, Natasha said. He kept developing in the incubator on damp towels and turned out fine. One of the painted turtles who hatched out of Mike's incubator spent twenty-four hours emerging from the egg. It took fourteen hours just to get his head out.

We're to move eggs with signs of hatching from the incubator to a bin with substrate. We need to make a little divot in the soil, with the opening in the egg facing up, so that the hatchling's hands will be able to grasp at the substrate and walk out of the shell. One of the eggs we lift out sports a paint's yellow-striped arm, reaching out to seize the day. Some of the hatchlings seem lethargic, struggling to escape their eggshells. One baby is wearing the back half of his egg like a diaper. If the fluid inside the egg dries before the infant is free, the eggshell can get stuck on the baby. Michaela shows

Matt how to fill a syringe with Ringer's and squirt it into the space between the turtle and its eggshell, then gently tease the baby free.

But one snapper hatchling can hardly wait. As I hold the egg in my hand, we watch, riveted, as the infant reaches one arm through the crack it has opened, then another arm. The baby rests for a moment. And then, in a spurt of amniotic fluid, the newborn—her shell domed high from being folded like a tortilla to fit inside the egg—launches her body from her calciferous confinement, ready to take on the world.

Like his miniature compatriots, Fire Chief, too, is eager to escape his confines. "He's had such an attitude since he started PT with you!" says Natasha. "He's so fired up, he clamors at the edge of his tank. I have to give him extra bananas to help him forget his impatience."

Fire Chief is so excited that his back claws slice Matt's knuckles as he's carried up the basement stairs and out to the Turtle Garden. He's raring to go. His neck sticks out like a periscope, and he immediately begins his march around the fence line.

His favorite spot to rest seems to be near the farthest edge of the enclosure, facing an unfinished outbuilding, beyond which lies a small wetland several acres away—which Fire Chief clearly knows about. Here, at rest, he takes enormous breaths that swell his throat and even the front of his legs. Cris had told us about this behavior at Turtle Survival Alliance; it's common to all land and freshwater turtles, and called "gular pumping." It is more than just breathing. Fire Chief is inhaling scents and flavors, literally filling himself up with the world around him.

Sometimes he tries to climb the fence, his great claws scraping until he is standing up on his hind legs—and then, to our dismay, he tips over onto his back. "He looks empty when he's upside down," says Matt. "A turtle upside down is like seeing a hermit crab out of its shell. It's shocking." We rush to flip him upright. As he recovers from his momentary upset, Matt and I stroke his head, his neck, his arms. Soon, he is on the move again.

Other turtles now join us for Fire Chief's physical therapy sessions. Tax Man, so named because he came in April 15, has a head injury from a car strike; he doesn't move for an hour. Another, a turtle between ten and fifteen years old, still called Number 30, has slimy moss on his back, is very fat, and snaps and lunges the minute he's set down. ("He defends himself very well!" boasts Natasha.)

Snowball joins often, occasionally with her tank-mate, Special. At first, Snowball didn't move much. But soon it's clear she particularly enjoys crawling through the tunnels. In one session, she walked the full length of the two longest walls nine times in half an hour.

"Fire Chief is definitely making progress," Natasha observes. As the weeks pass, it seems that more and more, he is holding up his plastron, standing tall on all four of his increasingly strong legs.

"He's filling me with hope," Natasha tells us one day. "The wild— that's his home without a doubt. He could be king of his pond again!"

Alexxia is more sanguine about Fire Chief's prospects. "I don't know if he can be released back into the wild," she says, "ever. But there's no doubt he is doing better." On the days she must work late fixing appliances, she comes back to TRL exhausted, but perks up after a few minutes in the Turtle Garden. She's especially proud and surprised to see Snowball doing so well.

"When she came in, she was so bad off, everyone wanted to euthanize her. And once, she drowned in her tub overnight and died on me!" she reminds us. "I pumped her full of air and got her back. And anyone else would have taken off that left front foot. I fixed you," she says to the turtle, "didn't I, baby pie? But," she says, turning to us, "it's still surprising to me to see this past year's momentum. You still kinda stupid," she says, turning back to Snowball, "but you're OK!"

We did not know at the time that Natasha and Alexxia were doubly surprised. The first time Alexxia saw Matt and me stroking Fire Chief, she almost couldn't believe it. "They're *touching* him on his *head*," Alexxia had whispered into Natasha's ear, so softly that we couldn't hear. "Did you tell them they could do that?"

"I didn't!" replied Natasha. "I thought *you* must have!"

Many sessions later, when Matt and I learned, we laughed. "Nobody said we could touch him," Matt said. "Fire Chief told us."

"It doesn't seem like ninety days since we were last at Teamsters," says Michaela. But it's been longer, because that's how long it takes for snapper eggs to hatch—and now, on August 18, we're back on the Teamster grounds, bearing a shallow bin of dirt crawling with seventy-five babies. All of them hatched from eggs collected on various expeditions to the edge of the parking lot. Because painted turtles often overwinter in the nest, the League will keep those babies and let them grow to release them in the spring, when they're stronger, bigger, and less likely to be eaten. But we're releasing the baby snappers today, and the maintenance man who alerted us to the nests, Scott, is meeting us here to take part.

His mask can't contain his smile as we all sit on the grass at the edge of the parking lot, and I remove the lid to the heavy plastic bin so he can peer inside. "There's *seventy-five* of them? That's unbelievable!"

Just hours after they hatch, baby snappers' tall, egg-hunched shells flatten out. The inch-and-a-half-long infants look like perfect miniatures of adults—complete with minuscule tubercles on their legs and arms, and tiny osteoderms on their long tails. That may be why almost everyone loves baby turtles: Born wrinkled and bumpy, they look like baby grandmas and grandpas. The incongruity of ancient-looking infants makes us laugh with delight.

"And they're able to go right into the water and right back in nature?" asks Scott.

"Yes indeed," assures Natasha. "These little guys know everything they need to know already. They're born with knowledge going back to the dinosaurs."

These babies are also born with their first meal. Michaela points out their "rubber belly button," a yellowish bump on the plastron, a yolk sac from the egg, which provides the baby with nourishment until it is absorbed into the body. (In some species, like paints, the yolk sac can be an alarming red color and as big as a grape, making it look like the baby has a horrible hernia or tumor.) Natasha likes to joke that it's as if these turtles' moms have packed their lunch for them.

"Here," Michaela offers Scott, "want to hold a baby?"

Of course he does. He stretches out his palm—and has to curl his fingers shut to keep the little turtle from racing off.

"How long do they live?" Scott asks.

"If everything goes perfectly," answers Natasha, "they can live one hundred and seventy-five years."

Scott's brown eyes widen in astonishment. "That's amazing!" he says.

Natasha then outlines the plan: We'll release the babies according to the "five by five" rule: five baby turtles can be released together, but the next five must be released no closer than five yards away. "There are advantages to small groups," explains Natasha, "but you don't want an all-you-can-eat turtle buffet for a predator."

We carry the box with us as we walk up the slope from the parking lot, then down through a ferny forest. Soon we arrive at the shimmering wetland. Birdsong mingles with truck noise. The sun shines almost merrily. Butterflies and dragonflies flutter above the pond, water skimmers skate across its surface—beneath which, nurtured in the cool, dark water, at least some of these babies may grow as big as Fire Chief and survive long after all the humans present today are gone.

"OK," says Natasha. "Everyone, take five turtles."

Michaela counts out five turtles for each of us, and we disperse to accomplish our momentous deeds. Here, among the swords of wild iris, the lacy scrim of duckweed, the flat pads of the water lilies, and the tall purple flowers of pickerel weed, we're repairing an ancient compact, one that was broken by cars and asphalt and concrete.

For each of us, this is a sacred moment. Some of us place our turtles in an inch or more of water and let them swim away. Some release the babies at the water's edge and let them crawl in themselves. I like to let them walk off my palm into the shallows.

"Good luck to you, buddy!" Scott says to his first one. "Wishing you one hundred and seventy-five years of happy life!"

It takes us only fifteen minutes to release all the babies. "I'm just a plumber," Scott says humbly, "but I've never seen anything like this. You're an honorable group of people. I've never known anybody who does stuff like this."

"You care. And you found us," says Natasha softly.

"It's a big role you played," Matt reminds Scott.

"In the world we live in today," Scott says, "what else can you do to make such a difference? What we're doing today will live on long after we're gone. One hundred and seventy-five years!!"

There is a big caveat, of course, to that projection: "If everything goes perfectly." As we have noticed—amid racial violence, global pandemic, the world on fire, the sea polluted, the climate deranged—everything does *not* always go perfectly.

Even for the babies who beat the odds, and grow too large to be eaten by frogs, or fish, or birds, or raccoons, or minks, or skunks, or foxes, or dogs, there are new gauntlets to face at each turn. Even if they grow as old and mighty as Fire Chief, the dangers never end. "In sixty, seventy, maybe one hundred years," Matt had mused to me at one of our therapy sessions with the Chief, "he had one bad day. That's all it took to nearly wipe it all away."

But meanwhile, what we know in this moment is that, on a

perfect August day in this glorious New England wetland, we have helped to restore seventy-five perfect, healthy baby turtles to their birthright in the wild. "It's a great thing!" Scott practically crows as we return to the parking lot. "How many things can make you feel this good?"

Just a few days later, we are back at the Torrington nesting grounds, checking the nests. We join as often as we can. "It's like Christmas morning almost every time we go out," says Emily. The Turtle Ladies check each nest several times each day, so that hatchlings don't bake in the sun. This year, for the first time, we have installed lids to protect the hatchlings from hungry crows, and Emily sewed dozens of "curtains" to shade the nestlings from the hot sun. We lift off the lids, pull the curtains aside, and there they are: perfect little replicas of the adult snappers, paints, wood turtles, Blanding's turtles, spotteds.

I love to bring children to help transport the babies from their nests to the river. One day I invite a visiting five-year-old and her grandmother. The little girl is fascinated with every aspect of the landscape, from the turtles to strange, sea-creature-like fungus known as earth stars, to the miniature flowers of purple love grass and sand joint weed. Another time, I bring a thirteen-year-old neighbor, Hannah. That day Matt hears scrambling, tiny toenails scraping on dirt. Wood turtle babies are hatching! Hannah and I suck in our breath. Four have emerged and are trying to climb the edges of the nest protector, while three others are erupting from the earth. "I remember seeing their mother doing the turtle dance, digging that nest," says Emily. "Oh, what a wonderful morning!"

My twelve-year-old friend Heidi accompanies us again and again. One day she arrives with her dad, a policeman, in tow. We revisit a snapper nest we had named after her—a nest she helped protect—and learn with delight that it's already hatched out twenty-six baby snappers. Other nests have more to come. We check on one that began yielding babies last night and which Emily assures us isn't done yet.

Heidi lifts the screen covering the nest protector to discover— "Oh my goodness!"—three babies. "No—four!" she cries, as, at Emily's urging, she gently brushes away loose soil around what is obviously an exit hole. "No—five! Wait, six—oh, wow, no, there's eight!" The babies seem sleepy when we lift them from the earth into the shallow water and sand in the red plastic bucket Jeanne and Emily call "the red bus"—but by the time we make it to the river, they are scrambling, eager, as if the water has called them to life. Powering back up the steep, sandy slope, Heidi gets tangled by a wild rose, its thorn leaving a cut on her finger. "It's proof we were here," says her dad, and this leaves her pleased with her small injury. Otherwise, Heidi agrees, "today might seem like a dream."

By Labor Day, the hatching season is nearly over. By September 12, the Torrington nesting grounds will have yielded a new record: 712 baby turtles hatched, including 528 snappers, 4 spotted turtles, 52 paints, and the largest number ever of endangered Blanding's (33) and wood turtles (96).

After checking the nests, when there's time, we sit together on the ridge again, looking down on the steep slope the turtle mothers must climb. Once, Emily tells us, she watched a mother snapper make it almost to the top—then tumble back down. She started right back up again.

Depending on the day and who has shown up, the humans gathered here might range in age from tots—practically hatchlings—to sexagenarians (which puts Emily and me in the class of elders with Fire Chief). We're all in this together. With the nesting grounds to our backs and the shining river before us, we are witnesses poised between potential and fulfillment—confident that once again, the world is being made new.

9.

Waiting

Fire Chief

"He's *so* handsome!" exclaims Jeanne, bringing her hands to her chest as if to contain her throbbing heart. On a cool day in September, she's come along to TRL and is meeting Fire Chief for the first time.

Matt and I had already shown her and Emily dozens of photos and videos of Fire Chief, which are filling up our cell phones and increasingly dominating our social media accounts. We take the videos to compare his progress walking as he continues his physical therapy. But Matt and I both find ourselves watching them on days we can't come in, just because we miss him.

Now that Jeanne sees the great snapper in person, she's even more impressed. His size is more massive, his head more mighty, his neck more majestic, his gaze more mesmerizing than any of our images could capture. And though it's cold enough that we humans have donned wool sweaters for our therapy session today in the Turtle Garden, Fire Chief's shell is still warm from his 78-degree tank, and he's remarkably active.

He takes off from the spot where I place him by Mount Olympus, and heads to the fence. He turns left, walks through a pile of fallen leaves, and marches determinedly up a gentle slope. He U-turns halfway up, then heads back downhill. He hugs the fence line that runs nearest the wetland. He often stops to open and close his beak, drinking in the tastes and scents of the water, the plants, the fish, and the other turtles living there.

Snowball and Special have joined Fire Chief outside today. Other than defensively lifting her back end high, displaying the sharp scutes at the rear of her shell, Special is largely immobile. But Snowball is active. At the opposite end of the pen from Fire Chief, she crosses through her favorite tunnel. Her head tilt is almost gone now, and she moves with awareness and purpose.

Now Fire Chief makes a turn and corners the fence, marching toward Snowball's tunnel. Head out, neck stretched long, he advances toward the much smaller female. They are on a collision course.

The two snappers are four inches apart, and I'm afraid he's going to stomp on her. But she's not: She boldly takes a step toward him. The two turtles pause, and Fire Chief opens his mouth once, twice, thrice.

"He's chemming messages to her," explains Natasha. "He's chemically receiving and probably transmitting information."

Snapping turtles may mate from May to September; the season is flexible, as females are capable of storing sperm, in some cases for years. Is this a sign of romantic interest?

Jeanne does the voice-over for Fire Chief to Snowball: "Do you want to have my babies?"

I burst out spontaneously, "Hey, *I'd* lay those eggs!" and we laugh.

A little while back, a girlfriend my age was gushing over a handsome Olympic swimmer she secretly fancied. "But he's half my age. And he's married," she admitted.

"Well, I've got a crush going on, too," I countered. "He's strong, he's gorgeous, he's age appropriate—*and* he's single."

"Oooh," urged my friend, "do tell!"

"But," I confessed, "he's a reptile."

Snowball pulls in her head and stretches up high on all four legs, so she looks like a footstool. She takes a step backwards. Fire Chief also stretches up.

Snowball walks forward, climbing over Fire Chief's head, over his massive shell, stepping on his long tail—and then the two snappers go their separate ways. Apparently, she wasn't thinking of our handsome friend *that* way.

But still, Snowball and Fire Chief's behavior signals they are aware of a change. It could simply be they are more social now because both are feeling better. Or it could be they are more active now because they sense the season changing.

Sometimes, watching over three famously slow reptiles, even if one of them remains almost immobile, can be surprisingly challenging.

On a day in early October, while Alexxia is doing paperwork, Michaela is cleaning box turtle habitats, and Natasha and Matt are busy with another project, I am once again in the Turtle Garden, supervising physical therapy for Snowball, a snapper named Stockings, and Fire Chief.

Stockings, slowly recovering from head trauma, sits still as a stone. But the moment Fire Chief and Snowball arrive in the Garden, they disperse in opposite directions. While Snowball trundles toward her favorite tunnel, Fire Chief stomps off immediately toward the water. But he doesn't stay long. Next he heads rapidly down the steepest side of Mount Olympus. I hover over to catch him in case he begins to flip.

Snowball, meanwhile, is traversing a cement step in front of a shed that forms one of the borders of the six-sided Garden. I rush over to make sure she doesn't face-plant.

Snowball crosses the step successfully and then rests, considering her next move—which is, again, to head back into her favorite tunnel. While Stockings sits unmoving, Fire Chief has made it safely down Mount Olympus and has reached the lowest corner of the pen. Fallen oak leaves cushion his plastron here, but they also make it more difficult for his claws to gain purchase. It's a real workout for him. Though all four of his legs are moving, he is not consistently holding up the back of his plastron. He waddles from side to side, and I can see where the sharp edge of his carapace is landing on his left side with every other step, hitting one toe of the back left foot hard enough to carve a shallow cut.

He's been cutting himself like this each time he's been out for the past several weeks. The first time it happened, we hoped it was just a random accident. Then we had hoped that as his muscles revived, he would stand tall and the problem would stop. But the same injuries reappear week after week after only an hour or two of therapy. The wounds are small and superficial, but still worrisome: a message written in red blood that his back legs might never completely recover.

I glance back to check on Snowball and Stockings. Stockings hasn't moved. Snowball has emerged from the tunnel and now heads toward Fire Chief.

At the approach of the much larger turtle, Snowball turns back. Fire Chief plows ahead. His neck is stretched out as he pauses, filling his throat with the chemical messages from the nearby wetland. He wants out of the pen. Then suddenly, it appears he has an idea. One short section of the fence here is covered in decorative latticework. He takes two steps toward the wall, and using his massive front claws, rises up on his back legs and attempts to use the latticework as a ladder to make his escape.

For most of New England's wild turtle species, autumn is a time of great urgency. Since reptiles are ectotherms—cold-blooded, with body temperatures matching the surroundings—they must find a safe place to spend the cold months, an area with waters deep enough not to freeze solid, and where predators will not disturb them during brumation. Studies have found that many snappers travel outside their summertime home ranges to find the right spot, and they often (though not always) return to the same area each fall. Ever since he was a hatchling, for almost all of his more than half a century on this earth, Fire Chief has heeded this compelling call; surely he feels it now.

Fire Chief's imposing claws make a ripping sound on the wood. I'm awed by the strength of his front legs. I'm humbled by his determination and focus. But his rear legs can't hold him up, and now all his weight is coming down on his back feet, the sharp scutes of the back end of the carapace digging into the skin of his toes.

I pull him away. Though he tries to hang on tight, he doesn't attempt to bite or claw me. When I place him flat on the ground, he sits almost sullenly, his head halfway drawn in, facing the forbidden latticework. "I'm so sorry!" I tell him, stroking his head, his hands, his feet, his tail. "You have to be patient. I swear to you, we will do everything we can to get you healed and back to the wetlands. We won't let you down."

He sits still for two minutes, gular pumping. It's as if he's considering his options. He comes, at last, to a decision: He's going to make a second attempt to scale the fence.

Head thrust out, his great neck straining, he grabs the latticework and starts to climb. I am holding up his shell so he doesn't further cut his back feet, but find I am, unfortunately, assisting his ascent. I try to pull him from the lattice, but he won't let go. I need both my hands to hold his weight; I cannot free his claws from the wood. If he gets to the top and crawls over, my arms aren't long enough to prevent him from crashing to the ground.

What am I going to do?

Matt appears like a guardian angel. He holds Fire Chief while I try to dislodge his claws. But I need both my hands to free just one of his, and the minute I get one hand loose, he clings harder with the other. Michaela also comes over—and with three of us working together, we finally get the giant turtle off the wall.

Matt holds him up so we can examine the plastron. It's not bleeding, but it's been scraped. Blood oozes from a small laceration on his left bridge, where the plastron and carapace join. And in addition to the wounds on his toes—they are like paper cuts, and surely hurt us worse than they do him—now he also has a new cut on his right rear leg from the weight of the sharp scutes of his carapace.

What can we do to protect him from injury as he strengthens his legs? Can we cushion his back scutes somehow? Can we construct some sort of shield for his plastron? "Maybe we need to build some kind of walker for him," suggests Matt.

My strong human friend dons gloves and carries the protesting turtle back to his tank. Besides, we have an urgent errand to attend to this afternoon as well—and it also was brought on by the approach of winter.

Yesterday, Natasha had started digging the hole. She dug for four hours. The ground was so parched, she said, it felt like digging in concrete. Today, Matt spent an hour helping her finish, creating a chamber three feet deep, three feet wide, and three feet long.

The search for a hibernation pond triggers a second turtle migration—and more patients and casualties to TRL. Surely this is the quest that set Number 111, an enormous sixty-pound male snapper even bigger than Fire Chief, probably a century old, across a road and into the path of a car seventy-two hours ago.

But unlike the Chief, he did not survive his injuries.

And that is why Alexxia and Natasha have scheduled an emergency mass burial for this afternoon. TRL's five-cubic-foot mortuary freezer is nearly full. Number 111 is too big to fit. His corpse is starting to smell.

We begin unloading the freezer at four p.m. We all wear plastic gloves. The expired patients are encased in plastic freezer bags of various sizes. Michaela reaches into the freezer and hands bags of frozen corpses to each of us. We tear open each bag and remove the carcass, dispose of the bag and any bandages or tape into the medical waste trash can, and place the bodies into a big bin, as black as a hearse.

Some of the turtles look perfect—except for the fact that they are dead. Others have metal plates on their shells, or silver tape. Some are smashed to bits. Natasha urges us not to linger so we can get the job done. "We'll pause and reflect on them at the grave site," she says.

Alexxia, Natasha, and Michaela are always respectful to the bodies of the patients they have lost; I was deeply moved when, on a different day, I had watched Michaela remove the eggs from a mother turtle who had died, and place the eggs in our incubators. It was a long, delicate, fiddly operation. Why not simply remove the plastron? I asked. "I try not to open them more than I have to," she'd told me. "It's a respect thing for me. And when I'm done, I try to put her back together as naturally as possible."

But turtles aren't supposed to look like this. "It's weird seeing them frozen," says Matt. Sometimes the bags are stuck together and we have to pry them apart with a screwdriver. Sometimes a patient was still oozing blood when placed in the freezer and the fluid has frozen to the plastic, making the corpse hard to dislodge. Sometimes, someone's claws are stuck in the seam of the plastic bag, as if they are still trying to cling to this side of life.

There are also bags filled with unhatched eggs from the summer. Unborn babies, turtles big and small—it feels terrible to be dumping them all into the dark bin. But there are so many, we must work fast. Otherwise we will still be shoveling dirt into the grave at dark.

We complete our grim work quickly. Natasha and Alexxia bring the bin containing Number 111 out to the side yard, and Matt and I carry the other.

The grave awaits under a young walnut tree. Alexxia wordlessly picks up Number 111 and places him in the ground first. "Sorry, kid," Natasha whispers to him. Next, each of us reaches into the deep black bin to pick up one patient at a time, to lay each, tenderly, in the earth—right-side up, because even in death, we cannot imagine any turtle wanting to be upside down.

Matt and I recognize many of these patients: Tacos, the feisty female paint who came in after a dog chewed her shell. Number 34, the snapper the animal control officer brought in from Ashland Avenue one of our first days as volunteers. Willow, a beautiful box turtle who had been kidnapped from the wild, held as a pet for decades, then relinquished. She had been housed with another female boxie, Cottonwood, and when Willow died—of causes unknown—Cottonwood had been lethargic for weeks afterward.

Each one of these turtles had a story, some long, some far too short. Seeing each of them as corpses hurts in a different way. I pick up a tiny baby in his egg. Just his head, and one hand, made it out. He lived only minutes.

Matt picks up a turtle whose dark shell is dotted with yellow specks. "Oh!" says Matt, as if physically hurt by the sight. "A spotted turtle!" This one life snuffed is a significant loss to an endangered population.

Natasha holds a wood turtle so flattened she looks like she was run over by a steamroller. "I can't imagine why anyone even brought this one in," she said. "Not even one egg could have survived."

In her two gloved hands, Michaela holds Curio, an adult female painted turtle who had arrived at TRL on her birthday, June 5. Michaela had spent dozens of hours with her, speaking to her, holding her, and often just breathing with her. "I have deep conversations with turtles, deep breathing with them, exchanging breath," she would later tell me. "I try to communicate to them in the best way I can that I am here to help them, not hurt them." Despite a large, Y-shaped crack in her carapace and injuries to her head and bridge, Curio was remarkably curious and confident—until she was discovered dead in her hospital box on August 26. "There we go," Michaela says, her voice full of the calm warmth with which she always spoke to her friend.

And then we turn to lift up the next turtle, a big snapper.

It's Skidplate—he fought to live so bravely! Week after week, he beat the odds, tolerating the treatments, while Natasha held one of his front feet to help him bear the pain of the injections meant to save his life. We all thought he would make it.

"Death seems to steal the dignity from such a dignified animal," says Natasha.

Finally, all the frozen turtles are in the grave, thirty-two in all. "This is a hole full of turtles we tried to do our best for," says Natasha. "This is the first and only time you will see them all together. We tried with every one of them."

Michaela silently weeps. Alexxia puts her arm around her. "You took them further than they would go otherwise," she says.

"You should be proud of those you saved," Matt tells her gently.

Natasha pronounces a eulogy: "From an adult one hundred years old, to tiny babies," she says over the grave, "they all lived, even a little bit. They all tasted life. They all felt the warmth from Mother Earth. Some defied the odds to become monster turtles. From Mother Earth they crawled, and to Mother Earth they return. Let's complete their final nest in the ground by securing them. Everyone, feel free to place rocks over them."

We cover the bodies with rocks for a practical reason: They act as a barrier to digging animals who might unearth the corpses for a meal. We certainly don't want the dead to be disturbed, but we also don't want to draw predators to the property, so close to the Turtle Garden where rehab patients might be exercising, or to the nearby wetland where wild turtles live. But as we place these weights, I am reminded of the ancient Jewish custom of leaving pebbles or stones on a grave.

The origin of the ritual is unclear, but it might be for roughly the same reason we're doing it: to keep demons, or other unwelcome visitors, away. An alternate explanation is the rocks anchor the soul in the earth, so it stays with us always. Some say the custom dates to the times of the Temple in Jerusalem, when Jews marked graves with piles of rocks to warn certain priests away—such priests, known as *kohanim*, tended to offerings, and were said to become ritually impure if they came within four feet of the dead. Others say that stones are just better than flowers, because they symbolize the permanence of our memories, and stand to remind subsequent visitors that the departed one is not forgotten.

No matter the origin, Jews all affirm it is a mitzvah—a good deed, an act of kindness and empathy, performed to fulfill a commandment of Jewish law—to place stones on the graves of the dead. In Hasidic teaching, the word *mitzvah* is said to derive from the root word *tzauta,* which means "connection." So each rock we place in the grave is a reminder: Those of us still living and those who have lived before us are still connected, our bond unbroken by death.

We each take turns throwing shovelfuls of earth on the grave. "What do you think happens after we die?" I ask Natasha. "I think we're freed from our body to become the original spark," she answers, "like a firefly flying in the night. We're returned to Mother Nature, the source. We're all made of the same soul stuff. I can't help but see that with these old turtles. These big snappers know how to make their brightness breathtaking," she says. "We lose a lot when we give up. But turtles never give up. And we never give up on a turtle."

I ask Alexxia the same question. She looks me straight in the face with an expression as surprised as if I just asked her to explain the role neutrinos play in particle physics. "I dunno," she says. Then she turns from the grave and strides back to the house, where hundreds of living turtles need her attention.

. . .

In the third act of the Thornton Wilder's play *Our Town*, the stage is set with three rows of chairs, representing the graves of the spirits sitting in them. The youthful protagonist, Emily, has just died in childbirth, joining others—including her mother-in-law, Mrs. Gibbs, and Simon Stimson, who hanged himself in his attic—who are already there. Those who'd arrived earlier sit patiently, calmly, staring dispassionately down at the life of their town, Grover's Corners, just . . . waiting.

Here and now, in the world of the living, the humans are all waiting—waiting for something to happen that never seems to come. No summer vacation. No back-to-school in the fall. The pandemic wears on; Black Lives Matter protesters gather day after day; and as their nations burn, Brazilian president Jair Bolsanaro and U.S. president Donald Trump continue to deny the overwhelming scientific evidence proving human-caused climate change. The seconds and minutes and days are exactly the same length as they were last year, but now they seem to drag.

Time spent waiting feels different to different people. The German psychologist Marc Wittman asked study subjects to sit in a room with nothing to do and estimate how long they spent there. The actual experiment lasted seven and a half minutes. Some folks thought the duration was only two and a half minutes. Many said it felt like twenty.

Perhaps the longest day of the fall that year was Election Day, 2020—which literally lasted not just one day, but four, before the results were called. Matt and I came in exhausted from listening to election coverage all night but even more so from the uncertainty. With mail-in votes uncounted, it's unclear which of the candidates have won the presidency. On the drive to Southbridge, listening to the news on the car radio, the tension was so great that we would simultaneously just scream.

Sprockets greeted us at the door wearing Alexxia's "I voted"

sticker on his shell, then went back to enjoying a heaping breakfast of greens, watermelon, squash, and strawberries. Pizza Man
stomped over to be stroked. We had cause for celebration: Apricot,
the elongated tortoise whom Matt especially loves, had, with the
help of a shot of oxytocin, finally delivered the four giant eggs we
had seen on X-rays. We had feared the eggs were stuck. We calmed
ourselves by changing filters and scrubbing and disinfecting the
slate slabs on which the box turtles' meals were served. We took five
snappers outside in the 63-degree sunshine: Snowball, Steampunk,
Spunky, Special, and of course, Fire Chief. To protect the toes of
his weak back feet from the sharp edges of his shell, we cushioned
the pointy scutes by wrapping the entire edge of the shell with blue
vet wrap, which is like a sticky ACE bandage. The tape is decorated
with type that reads "Stud Muffin."

Every one of the snappers is on the move: In the Turtle Garden,
Spunky turns in a circle and hisses in response to Michaela scratch-
ing her shell. Special walks fifteen feet for the first time. Snowball
leaves the November sunshine for the icy-cold water of the little
pond, heeding the wisdom of the ancestors urging her to find a
hibernation pond *now*.

Matt and I take Fire Chief to a grassy area out front with fewer
rocks to scrape his plastron. He heads purposefully toward a patch of
goldenrod and multiflora rose. Today his back legs look weaker, and
his left leg in particular seems to drag more like a flipper than a foot.

But happily, the vet wrap is working, cushioning his back feet
from the sharp scutes of his shell. Soon he seems to recover some
of his strength and holds his plastron higher. He picks up speed.

He too seems propelled by turtle wisdom, and feels called to the
season's errand. When he stops, he fills himself with gulps of air,
inhaling the world and its chemical news. Unlike the frustrating reports we hear on our car radio, this is news Fire Chief understands
and accepts, and to which he knows, without question, the proper
response.

Matt and I wish we knew the results of the election. But we are, in these moments, at peace with time. We no longer want to scream.

"To everything there is a season," reads my favorite passage from Ecclesiastes, the first verses of chapter 3, "and a time to every purpose under heaven." There is a time to be born, and a time to die; a time to heal and a time to mourn. There is a time to hurry and a time to wait. But waiting, for our kind, can be excruciating.

When I watched *Our Town* for the first time, the waiting struck me as the most distressing part of being dead. The calm of the older spirits confounded me. How could they bear the wait—especially since there was no end in sight?

But turtles are champions at waiting. In the North, they practice it each winter when they enter, and sustain for months, a state of semi-suspended animation.

The state of brumation allows turtles to cheat death by emulating it. They don't eat. They don't breathe. The heart may beat only once every few minutes. Metabolism may decrease by a full ninety-nine percent. Some species, including baby western painted turtles, can even freeze solid and still survive.

Most brumating turtles spend the winter at temperatures just above freezing. (Owners are sometimes advised to overwinter their box turtles in a container of dirt in a spare refrigerator.) Russian tortoises, who survive in the icy steppes of Central Asia, may construct burrows six feet deep for the purpose. Spotted turtles overwinter sheltered within grass tussocks, buried under ledges, wedged in among roots and cushioned by soft moss. Most of North America's aquatic turtles who overwinter do so in ponds and streams—sometimes leaving food-rich summer ponds for quieter ones, sometimes moving to a different part of the same wetland. In small waterways that flow all winter, snappers brumate buried under logs; in lakes, they wedge themselves beneath logs and stumps close to shore; and in marshy areas, they bury in deep mud. They sometimes overwinter in muskrat lodges; sometimes several snappers even brumate together.

Submerged in water, or buried in mud, beneath a body of water that might be capped with ice for more than a hundred days, snappers and other aquatic turtles have access to very little oxygen. No matter: They can breathe through their butts. They don't need their lungs at all, but absorb oxygen through the blood vessels close to the skin's surface, of which one of the richest areas is the cloaca.

To keep their cells alive in the absence of food, brumating turtles use energy stored in the liver and muscles. After months, this takes a toll and produces an excess of lactic acid—the reason our muscles cramp after too much exercise—which can be fatal. To neutralize the acid, turtles mobilize the calcium from their shells, the way you might take an antacid.

Though during brumation, a turtle's metabolism is completely transformed—you might mistake it for a turtle who is sleeping, or a turtle who is dead—a brumating turtle is not unconscious. It may not always even be immobile. Standing on the ice of a frozen pond, you might see a painted or snapping turtle swimming beneath the ice. Even if the turtle looks completely inactive, it is awake. In a study published in 2013, Danish researchers placed electrodes in the brains of red-eared turtles brumating, unmoving, in cold, oxygen-poor water. The turtles' brains responded to changes in both light and temperature. "Hibernating turtles are not comatose," the authors wrote, "but remain vigilant during overwintering."

Some species of turtles can exist in this condition for more than 217 days and still recover. How, I wonder, do turtles experience the passage of time during this state? Is it boring? Is it restful? Does it even feel like waiting at all?

Of course we can't know; we can't truly understand what any experience feels like for another person, not even our spouses, our lovers, our children—much less different species, creatures so different from us naturalist Henry Beston called them "other nations."

Humans, as we have seen, are notoriously bad at correctly perceiving the passage of time. Most people can accurately keep track of time for only five seconds, Wittman found. (And it takes about half a second for our brains to realize anything is happening at all, depending on which area of the body, and which sense, is bringing the news: Signals from the feet take longer to reach the brain than signals from the lips. We are all, therefore, living slightly in the past.)

Human frailties of vision, hearing, and touch limit our perception of time, Nicholas P. Money, a professor of biology at Miami University, writes in *Nature Fast and Nature Slow*. "Time is missed so easily, clocked [only] when our attention is on it," he states. "Even when we are watchful, however, we are aware of just a sliver of it."

Animals are "living by senses we have lost or never attained," as Henry Beston presciently noted in *The Outermost House*. (It was published in 1928, when scientists had hardly begun to discover the extent and variety of many animals' sensory powers—many of which are still unknown.) Many dogs, for instance, begin to wait by the door at a specific time each night for their owners to return. Some dogs will even meet their person's train—their timing often more punctual than the locomotive. (One, an Akita named Hachiko, continued to meet the habitual afternoon train each day for ten years—despite the fact that his owner never returned, having died unexpectedly at work.) How do they know what time it is? The Columbia University professor Alexandra Horowitz, a specialist in dog cognition, believes they can smell the passage of time. Scent molecules decay at a particular rate, which is why scent-tracking dogs can easily find the start of a trail and follow it to its end, not the other way around.

We cannot smell time. We humans receive up to eighty percent of our impressions, by one estimate, by sight. Yet we cannot see a flea jump. To us, the whirring wings of a hummingbird are just a blur. Why? Because "we live by seconds," Money reminds us: Our stomachs contract every twenty seconds, our intestines squeeze every five, our hearts beat about once a second. Stimuli in bursts of roughly two milliseconds completely escape our notice. We cannot hear the sound waves bats use for echolocation, produced at two hundred calls per second at eighty kilohertz. But the predatory larvae of certain winged insects, known in their youthful state as ant lions, can—and dive into sand to take cover when they do.

Insects can process far more images per second than we. A dragonfly watching TV would see over two hundred separate still images each second. Until high-definition TV, dogs, too, would see on the screen a series of still images separated by blackness, while we would see a smooth flow of action. The maximum speed at which a creature can see flashes of light before the light source is perceived as constant is called flicker fusion frequency, and it is one measure of how we and other animals experience time. A housefly's flicker fusion frequency is 250 flashes per second. A pigeon's, 100. A dog's, 80. A human's, 60. A sea turtle's—the only kind of turtle tested—is 15.

Based on the results of flicker fusion rates of thirty species of animals, an international collaboration led by scientists from Trinity College in Dublin concludes that the ability to perceive time is linked to the general pace of life. Why might this be? Because heightened visual processing takes a lot of energy. Published in the 2013 issue of *Animal Behavior*, the study shows creatures with small bodies and fast metabolic rates perceive more information in a unit of time—and experience time more slowly—than animals with slow metabolic rates. You would expect that an Etruscan shrew, whose heart thrums at a frenzied rate of twenty-five times each *second*, experiences each hour as lasting much longer than does a Galápagos tortoise, whose heart beats at a stately eight times a minute. And this seems only fair. The shrew may, with luck, live only two years; a Galápagos tortoise can live to at least one hundred and seventy-five, maybe older.

But one might hope that both animals, if allowed to naturally attain the end of their vastly different life spans, may experience a full measure of life. And if this is so, perhaps nature has granted the slow-living, long-lived turtles the patience of the calm spirits in *Our Town*. "Everybody in their bones knows that something is eternal," the Stage Manager tells the audience. The dead are at peace, knowing that at the end of their waiting, the eternal part of their selves, like a turtle in the spring, will emerge.

"We don't see, hear, or touch its passing," states *The Stanford Encyclopedia of Philosophy*'s entry "The Experience and Perception of Time," "and yet we would still notice the passing of time even if all our senses were prevented from functioning."

But that entry, last updated in 2019, did not incorporate a new discovery then not widely known: Along with the senses of touch, smell, hearing, and sight, we may have a sense of time, and special cells to detect and measure it. In the eyes of both sighted mice and mice bred to be born blind, the University of Oxford neuroscientist Russell Foster found a pigment known as melanopsin, which appears to respond to light—even if the light is unseen—in a way that links an organism's body clock to night and day. Information from these time-sensing cells travel along the optic nerve but pass the brain's visual cortex, where information from our retinas' rods and cones are processed. These signals continue to a completely different, deeper area within the hypothalamus. A paired clump of cells here known as the suprachiasmatic nuclei may be the area where our unconscious but essential circadian rhythms are coordinated. Melanopsin has also been found in the skin of frogs; other kinds of opsins have been found in the skin of squids and octopuses. The implications are mind-blowing: We might be able to sense time with our eyes, and other creatures may be able to see with their skin.

Scientists have hardly begun to investigate turtles' exquisite senses. Honed by time, turtles' faculties—including some perhaps as yet undiscovered—have been steering their lives reliably for longer than those of most other terrestrial vertebrates still found on this Earth. But sometimes this is not enough. To a turtle, a car whizzing past at sixty miles per hour may be a blur, no more visible to them than the individual wingbeats of a hummingbird are to us. Fire Chief, Snowball, Number 111, and so many others of our patients may never have even seen the cars that hit them.

Hundreds of millions of years of evolution did not prepare turtles for the convulsion of change that humans have wrought in the geological blink of an eye. For eons, hatchling sea turtles, so numerous they once blackened the beaches of Texas and Mexico, would ride the Sargasso Loop current to the middle of the Atlantic Ocean; in three to five years, the youngsters would swim north to summer in Cape Cod Bay to feed on its rich supplies of crabs, jellyfish, and algae. Each fall, before the water temperature dropped to 50 degrees Fahrenheit, they would head south to warmer waters. But climate change is now warming the waters of the Cape, and the entire Gulf of Maine to its north, faster than anywhere else in the world's oceans. Many turtles linger too long. By the time they leave for the migration, waters of the Atlantic are too cold for them to swim. They get trapped by the hook-shaped landmass of the Cape. They are stunned by the cold: too cold to think, too cold to move, and finally, too cold to survive. If the wind is blowing in the right direction, they are washed back up on shore like driftwood.

In winter of 1974, Robert Prescott, director of Massachusetts Audubon Society's Wellfleet Bay Wildlife Sanctuary, began rescuing the first cold-stunned turtles washing up on the Cape's beaches. Now each winter, there are hundreds, sometimes thousands.

Natasha tells us about the plight of these turtles when we visit on the second day of December. She's been following the weather on a website called MagicSeaweed.com. It predicts that in three days, winds coming shoreward from the northwest, cutting across Great Island and Dennis, will blow in excess of thirty-five miles per hour: conditions ripe for cold-stunned turtles to be carried helplessly back to shore.

Normally a network of hundreds of Audubon volunteers on the Cape mobilize to comb the beaches at such times to rescue these turtles and take them first to the makeshift sea turtle hospital at the Mass Audubon Wellfleet Bay Wildlife Sanctuary and finally to New England Aquarium's animal care facility for rehabilitation. But this is not a normal year. After the Thanksgiving holiday, Covid is now killing one American every forty seconds. Massachusetts's medical centers are out of beds for the sick and setting up field hospitals. For safety reasons, the usual gathering of volunteers has been canceled.

But we will be going on Sunday.

Painted turtles winter beneath the ice.

10.

Sea Turtle Rescue

Kemp's ridley sea turtle

By noon on Saturday, we looked out our windows into a snow globe. Winter storm warnings extended from central Massachusetts to northern Maine. The low-pressure system was centered over the Cape Saturday, drawing northeast winds measuring sixty miles per hour on the coastline. A foot of heavy, wet snow—the kind that brings trees and power lines down—was predicted in New Hampshire and central Massachusetts. Federal weather forecasters warned that the weekend nor'easter moving up the New England coast was creating "treacherous to even impossible travel conditions" and could possibly evolve into a "bomb cyclone"—two words that, taken separately, evoke trepidation, but combined captured the attention of friends who knew about our patrol. Several urged me not to go. One wrote me with this prayer:

> *Through the long night watches*
> *May Thine angels spread*

Their white wings above thee,
Watching round thy bed.

"Will be thinking about this very prayer for the sea turtles washing up on the beach on Sunday, awaiting our rescue . . ." I wrote back—just before our Internet went out.

We woke Sunday to discover that more than 200,000 New England households were without power—including Matt and Erin's. Maybe this was a good thing, Matt declared cheerfully on the phone that morning—his yellow rat snake, Ernie, had gotten out again earlier in the week, and perhaps the cold would draw him from his hiding place. They had a generator to power the turtles' heat lamps and the refrigerator, he explained. Then he and Erin went back to snow-blowing and moving the tree that had fallen across their driveway.

By ten thirty a.m., the sun was out, turning the snow-rimmed trees along the road to Matt's house into sparkling white cathedral arches. The world looked like a Christmas card. Except for the flights of the bold chickadees, everything was still, frozen. It seemed a very odd day to go search for reptiles.

We meet the rest of the patrol team in the freezing parking lot of a closed coffee shop in Sandwich, Massachusetts. "Welcome to Sea Turtle Patrol!" comes Alexxia's voice from behind her black mask. Her blue-green eyes and two wisps of her dark hair are the only parts of her head not completely obscured by the mask and the crown and ear flaps of her knitted ushanka hat.

Who else is here? Bundled and face-masked, everyone looks like a different bandito version of the Michelin Man. There are eleven of us, including me and Matt and Michaela. Somebody is Mike Henry, someone else is TRL volunteer Mike Webster, another person must be Dan Tracey—a legendary hero who, on an earlier expedition, had carried a four-foot-long dead tiger shark back to the Wellfleet Wildlife Sanctuary lab for autopsy, slung over his shoulder—and someone else is Michaela's partner, Andi. Also volunteering is a Scottish couple, friends of Mike Webster's, whose charming accent beneath their masks I cannot penetrate.

"Today, or rather tonight, will be a real adventure," Alexxia continues. "It's going to get cold. It's going to get dark. Where we're going, there will be no roads. There will be no shortcuts. There will be no lights. There will be no resupply. There will be no rescue."

"There will be no bathrooms," adds Natasha.

Generously, friends of mine—now snug in their winter home in upstate New York—have offered Matt and me their summer house in South Orleans, farther down the Cape, to stay overnight, and the team a chance to meet and use the plumbing before our sortie. We reconvene there.

"Today we're doing one of the hardest turtle rescues," Alexxia resumes. "Bring the map up on your phone, and you'll see Great Island is a little needle of a peninsula out in the middle of the bay. It gets weird out there.

"Walking in sand is not a real delight," she continues. "And it gets very cold. So don't walk until you're exhausted. Remember you have to walk all the way back with the wind blowing at you. I want it to be a turtle rescue, not a human rescue."

Natasha reiterates the importance of a good flashlight. Our cell phones won't provide enough light, she stresses, and because of the cold, our phones' batteries will go dead. We should keep our cell phones in an inner pocket of our coats, she advises, and save it for when you find a turtle, to report to Alexxia immediately . . . if there is any cell reception at all.

We'll split into two teams, Alexxia tells us:

One team will patrol the front side of Great Island, also known as the outer band. This route is a little shorter, but because of the consistency of the sand, more difficult. Matt and I, Alexxia and Natasha, and Michaela and Andi will take this one.

The other is known as the Wellfleet Graveyard, named for lots of skeletons of sea creatures that wash up on the beach. It's a nicer walk, she tells us, but with lots of obstacles to walk around. Mike Henry, Mike Webster, Dan Tracey, and the Scottish couple will take this patrol, and meet back up with us for the last few miles of beach walk as we return.

There is a chance we'll change plans, Alexxia says: Coordinators at the Wellfleet Wildlife Sanctuary may need a different section of beach patrolled. There have been multiple strandings already. Now, we've been asked to leave, and stay out, later than we expected. High tide is at three thirty, but we're to leave at four thirty—leaving us even less light and warmth than before.

"Try not to get wet," Natasha pleads. Wet feet are a prescription for hypothermia and frostbite. "We CANNOT go into the water. And remember, more turtles might be on the beach as we come back."

Alexxia outlines the rescue protocol: If you see a turtle, you can either take it with you or set it above the tide line for retrieval on the way back later. Phone in the find. If a person is tired, they can head back to the parking lot with the sea turtle. Cover it with dry seaweed—never sand—to protect against the chill of the wind and wandering predators. Remember to cover the head. The animal will still be able to breathe.

Stranded on the beach, the turtles are even more helpless than they are in the cold water. "The water is insulatory," explains Natasha. It's 50 degrees in the water today; but the air temperature reached a high of 36 and will probably dip to a low of 30. The turtles would be doomed in the water; but if we don't rescue them, they also face certain death on the beach. Some may well be dead by the time we find them.

"Every turtle's got to come off the beach, dead or alive," Alexxia tells us. "Remember, we're looking for one of the rarest sea turtles in the world." All seven of the world's sea turtles are threatened—by everything from climate change to discarded plastic bags (which they mistake for jellyfish)—but Kemp's are the most endangered because they are the species most commonly caught in fishermen's shrimp trawls, gill nets, long lines, traps, and dredges. Because of their rarity, each Kemp's lost is a disaster for the species. "And it's very possible we will find live sea turtles today," Alexxia says.

When we arrive at Jeremy's Point on Great Island Beach, a gorgeous sunset spills across the sky. Purple clouds with blood-orange underbellies hang over a low layer of yellow light, over silver sea, over gray-green dune grass. Matt and I rush to the top of a dune for a better view. The wind whips sand into my eyes. It's almost strong enough to blow me over.

In the Black Panther—Matt's compact pickup—Matt has brought his ice-fishing sled: four feet long and two and a half feet wide, with twelve-inch-tall sides. It's big enough to haul his canvas ice-fishing house and gear, along with his dog, Monte—and even carried his father. "We can haul lots of turtles!" he exclaims. Mike Henry has brought a smaller, lighter one—the size a child might use to slide down a hill. We don hats and gloves and the final outer layers of our winterwear, and unload our gear. Some of us brought walking sticks, and all of us carry backpacks filled with flashlights and extra dry socks and power bars and water. At four thirty we slide down the bank of a sand dune onto the beach, and the teams split up.

The remains of an orange sherbet sunset hovers above the pewter sea. Visibility is excellent. We don't need our flashlights. It's windy, but we are warmly dressed; the wind is at our backs. There is no snow on the beach. We aren't cold at all. Now it's clear why we waited to leave till an hour after high tide. The water is very close to the edge of the beach—at high tide there would be, in places, no beach to walk on.

The last glimmers of sun drench the sand banks with a glow that makes this place look as if lit from inside. Matt is thinking: Everything is so beautiful; surely nothing bad could happen here. Alexxia is looking out into the foaming ocean. She is thinking: There are *turtles* out in that! And my thoughts are a prayer to the sea: If your waves hold any cold-stunned turtles, please bring them to us so we can help.

There are plenty of items scattered on the beach, but a quick scan reveals they are only rocks, shells, and seaweed. Because we're looking at the sand, we often fail to heed the surf until it almost catches us. We scamper away, giggling like children playing on a summer's day. Except in winter, especially at night, getting caught by the surf could have grim consequences. Because water steals body heat twenty-five times faster than air, wet feet in winter can turn into a medical emergency fast.

We've only been walking for ten minutes before my feet are wet. The surf has come in and gathered in a wide depression in the sand. We use Matt's sled as a bridge across the deepest part, but while the other side looks solid, my hiking boots sink and I feel water soak in. Fortunately Matt bought me wool socks and foot-warming insoles. I pray they will hold up in water and last for the next three hours.

Darkness rises. We turn on our flashlights. Now everything on the beach looks like a turtle. Rocks. Clumps of seaweed. A dead seagull. Now we are not just scanning but searching. We prod clumps with our shoes, poke them with walking sticks. Nothing. But every lump must be investigated. If we miss one, a young sea turtle who should live thirty to fifty more years will die.

Alexxia's 11,000-lumens flashlight ignites the whole beach. We stop at a whitish mass. It's not a turtle, but what is it? A dead common eider—a large, heavy-bodied sea duck. The males, like this one, sport bright black, white, and green plumage. These ducks have been dying at alarming rates on the beaches here due to a virus first identified in 2007—a localized "duckidemic." Whether the virus was this bird's fate we can't know; he's too badly decomposed to bring back for necropsy. And though half an hour earlier, it seemed that nothing bad could happen here, now that we're trapped in the wind and surrounded by darkness, we're far more aware of the giant, invisible forces that make every living thing seem tiny and vulnerable.

Alexxia walks ahead of us, sweeping her light. A few minutes before five p.m., about two miles in, she veers sharply to the left, toward a rock.

The rock is moving.

As if swimming against quicksand, the sixteen-inch-long, teardrop-shaped rock inches up the beach. As we approach, we see it slowly, slowly swinging forward the right front flipper, and at the same time, the left rear . . . and then, a full four seconds later, the opposite pair of limbs begins to move, in ultra slo-mo, a toy whose batteries are almost run out.

The turtle, a Kemp's, has no visible injuries, its sleek, serrated, gray-green shell marred only by a single tiny mussel near the center of the back—incongruously standing straight up as if it's hailing a cab. "What are you doing there?" I ask the mollusk. "Don't you know that you're riding someone who is alive?"

But then, everything about this encounter seems unreal. You're not supposed to see a reptile in winter. You're not supposed to see a sea turtle on land. And even though we tried to prepare to find the impossible, all the videos we saw readied us to find a turtle who was half, if not entirely, dead.

"That it's moving is a very good sign," says Alexxia.

We're elated the Kemp's is so visibly alive. But why is the turtle trying to walk up the beach, away from the life-giving sea? Ten years or so from now, and only if this turtle is female, she might make this perilous journey, along with hundreds of others, to land to lay her eggs in the sand. Not here, but in Mexico. And not now, but in summertime. In winter, on the Cape, no good could come to a sea turtle on the beach. It's far colder here than in the 50-degree water—and with the wind chill, the air temperature feels even colder. "The water is insulatory," we'd been reminded. Of course the turtle had no way to know we'd be looking here. Had we not come, had we not seen her, this turtle was walking toward certain death.

Sea turtles who looked much like this one have been living in the world's oceans for sixty-five million years, since the end of the dinosaurs' reign. Why should the wisdom of the ancestors fail now? Why use the last of your strength to pull yourself up the beach? What is this turtle thinking?

I think I might know. On an expedition to the highlands of Papua New Guinea, at the end of a long day's hike at ten thousand feet, I was suffering from both altitude sickness and, as it began to rain, hypothermia. I left my group in order to discreetly vomit. Then, disoriented, I wandered away, into the trackless cloud forest—where the rain obscured my footprints as well as the sound of a calling human voice. Other members of the expedition noticed my absence and rescued me before I disappeared completely, or I would have certainly died. I had done this same thing once before, in my twenties, after surgery. I woke up from the anesthesia aware only that something was badly wrong, so I should try to get away. A nurse caught me crawling, turtle-like, on hands and knees down the tiled hallway. Looking back, I remember that though I felt completely alone in the vastness and inexplicability of my illness—so surrounded and filled with it that I felt cut off from all others, and from the very idea that anyone might help me—I was not panicked. I was possessed of a single, instinctive certainty: I must escape. Like this turtle, I methodically staggered forward. From an evolutionary perspective, it makes sense: If the situation is hopeless where you are, you stand a better chance going almost anywhere else.

We all gather dry seaweed to make the Kemp's a cozy nest at the back end of Matt's big sled. The seaweed is lacy and light and strung with shells, fishing tackle floats, and small rocks, like Christmas tree ornaments. We pad the bottom of the sled and snuggle our turtle into its embrace, then cover the head, shell, and flippers with a blanket of seaweed. The foot-tall sides of the sled will block the wind. When I open my pocket notebook to record the time, the wind tears the page from the spiral plastic binding. It's as if the waves and the sea were saying, *Forget all that. Clock time doesn't matter here. You are now on turtle time.*

Rescuing the turtle brings us all a surge of energy—like we just chugged Red Bull. Surprise, distress, delight, tenderness, and urgency tangle like discarded fishing gear. Matt and I later told each other that at that moment, we felt we could walk all night.

5:15: Michaela spots another, smaller turtle. "It was a rock, but with legs and a head, and it was moving!" she says in awe. "I couldn't believe it!" This turtle has a cartoon-like bite mark on the left rear of her upper carapace—a healed bite from a shark. She also has a superficial, bleeding cut on the left rear flipper. Exhausted and cold, unable to eat for days, her heart so slowed she can no longer swim, she's been battered and scraped by tumbling helplessly against rocks and shells as the tide pushed her to us.

"Put her in there with her snuggle buddy," says Alexxia tenderly. We can't tell if the turtle is female, but in the absence of evidence to the contrary, since Alexxia so closely identifies with every turtle, it's no wonder she automatically considers this youngster a female like herself. We tuck the turtle in among the seaweed and her new friend. I wonder how they feel about the company.

Another jolt of Red Bull–like energy sharpens my senses. I feel hyperalert. What's that? And that? Seaweed. Rocks. Driftwood. A buoy. Some kind of plastic box. A lobster trap. Discarded fishing rope. And now, a different shape: It's not a turtle, but worth investigating. We gather round it. It's the remains of a fox in the sand, coming apart like a worn bath mat.

Alexxia leads with her dazzling flashlight as Natasha walks closer to the ocean. Now the tide is ebbing, the beach is much wider. There's far more sand to scan. Michaela and Andi fan out to the drier sand farther from the surf. Matt and I are behind, as he pulls the increasingly heavy sled with its load of turtles and seaweed.

How are the others doing? We haven't yet heard from the graveyard team, and with spotty cell service, we probably won't. I check periodically to see if anyone in our group has wet feet; I have extra foot warmers in my backpack and they are successfully keeping my soaked socks warm. Does anyone want a power bar? Some almonds? I worry about Alexxia; because she is so slender, she easily gets cold. Natasha is super fit from running and biking, but it can't be easy for a blind person to be walking at night by the sea. Michaela and Andi are strong and young, but they've picked the most difficult part of the beach to walk on, where the surf hasn't soaked and packed the sand. It looks to me like Michaela might be stumbling. But she tells me she and Andi are fine. Or at least that's what I think I hear before the wind blows her voice away.

The spotlight reveals turtle number three. "Another live turtle!" Alexxia announces. It's about as big as the first one, with no visible injuries. "Sy, you can pick her up," she tells me. She feels much lighter than a land turtle of comparable size. I expect her skin to feel rubbery, like a dolphin's or a stingray's. But no, it's remarkably soft, like the skin under Fire Chief's arms. I'm amazed something so soft can survive being flung up on the beach by the surf.

Now Matt is really feeling the weight of the turtles he's dragging. Because of the depth to which the foot sinks in sand and the energy to extract it with each stride, walking on sand takes two to three times more effort than walking on a hard surface, and in addition to this, Matt is now pulling thirty pounds of turtles plus perhaps ten pounds of seaweed on his sled. Even with the wind, he's so hot he stops to take off the sweatshirt under his coveralls. I pull the sled for him a few dozen paces, and am secretly relieved when he insists on taking it back.

At this point, we are all privately wondering when the beach will run out and we can begin the long slog back. It certainly feels like we have walked more than four miles. Alexxia looks at her cell phone at the map and confirms this is so. "Look where we are," she says. The pin shows us well over half a mile offshore—in the water. "This is crazy. We're out in the middle of the ocean." As high tide receded, a sandbar has appeared. We are walking on it now. "Bonus miles," I say cheerfully. I wonder what time it will be when we and the turtles get back.

5:40: Alexxia spots turtle number four, right at the surf line— also alive and moving, but barely, and about the same size as the last one. Each turtle is still a shock to me. But will there be more? If so, can we even carry them all? Is this beach ever going to end?

The bonus mile yields a bonus turtle. At 5:55, the furthermost reaches of Natasha's flashlight reveal the end of the sandbar. But before we reach it, we discover turtle number five, with a smear of red blood at the point where the neck joins the body. It looks like a shallow scrape, but nobody wants to distress the turtle by examining the wound. She is alive, and that's all that matters. We tuck her in with the others. Matt drops the sled here, as we advance to the point where the land runs out.

We are greeted by an explosion of birds—gulls who were resting on the beach are upset by our approaching flashlights. "We made it to the end!!" cries Alexxia.

Applause for our oracle, Natasha, I say—who, having been carefully following the weather forecast, correctly predicted this night that the sea turtles would need our help! A hand for Alexxia, the lead turtle spotter, and for her great flashlight! A hand for Michaela, who found our second turtle! And a hand for Matt, for pulling all those turtles—and who is now going to pull them back!

The applause fades, and we turn off our flashlights for a moment and listen to the wind—howling now—feeling the darkness upon us like a hand. We look out, surrounded by water on three sides. The land where we now stand was, as Alexxia's map showed, just hours earlier, covered by water. The land was here, but we couldn't see it—the sandbar, like the turtles, was revealed to us by the sea. Such wonders feel biblical to me: like Jesus walking on the water during the storm, or the parting of the Red Sea for the Israelites as they fled the Egyptians. Our meeting these sea turtles on the freezing beach, and saving them from their certain deaths, feels like no less a revelation.

Ahead of us, a flotilla of eider ducks bob perfectly at peace on silver waves, throbbing like a poem. The ducks belong here. We do not. Here in the darkness and the wind and the cold at the edge of the sea, we light-loving, warmth-seeking mammals are out of our element—like the storm-tossed Kemp's huddled in the seaweed in Matt's sled.

We turn and, with Matt dragging the heavy sled, walk back toward our cars, facing into the wind.

II.

Coming Out

The Chief in his wheelchair

A week later, warm and rested, we're back in TRL's 80-degree basement. As we prepare breakfast for the box turtles—giant blackberries, cantaloupe, and greens, with a side of chicken, a meal you could get at a spa—Matt, Natasha, Alexxia, and I review the rescue's successful conclusion.

Though Matt never complained, he admits now that hauling the sled, eighty pounds once full, all those miles over the sand, with the wind in his face, was exhausting. Fortunately, we eventually reunited with the graveyard crew, who had found no turtles to rescue. They took over dragging the sled for the final two miles.

By eight p.m., all of us were bushed. Natasha was staggering, Alexxia, Michaela, and Andi were spent, and I could barely push hard enough against the wind to make it up the last, steep, sandy slope to the parking lot. But our accomplishment also left us exhilarated.

"We got five of the most endangered sea turtles in the world off the beach," Alexxia reminds us. "They definitely would have died. And now they definitely will survive." By now, the five turtles we rescued will have been transferred from the Wellfleet Wildlife Sanctuary, where we'd brought them that cold, dark night, to New England Aquarium's animal care facility in Quincy. There they will continue their recovery until their release in the spring.

Spring seems a long way off. At TRL, everything has changed. The incubators are empty. The healed and the hatchlings have been released. From possibly as many as a thousand turtles, we are now down to fewer than 250 remaining patients and permanent residents.

Still, there's seldom a dull moment. The box turtles are getting into mischief. When we deliver their breakfasts, we find Juniper is desperately trying to crawl into Acorn and Cherry's habitat, while Speedy is trying to scale the walls of his enclosure to get in with Walnut. A few years ago, Percy escaped his habitat to visit a number of his female neighbors. As a result, a beautiful, shy boxie named Patience soon laid eggs that hatched into seven healthy babies.

But such travels can be dangerous. People think of turtles as placid, but sometimes they fight, occasionally savagely. Turtles will argue over dominance, food, basking spots—and sometimes just because, for whatever reason, one takes a dislike to another. (A pair of Galápagos tortoises had lived together happily for 115 years— until, one day, the female, Bibi, suddenly couldn't stand the male, Poldi. She attacked him so viciously and consistently that the keepers at Austria's Reptilienzoo had to build separate enclosures for the two.) We quickly erect screens and lumber to heighten the walls between the habitats before anyone suffers a bloody bite.

Fire Chief is getting feisty, too. We learn, to our dismay, that he briefly flipped upside down in his stock tank last week, when his water was low. Natasha assures us he was in no danger of drowning, and insists this is actually a good sign: "He's getting better, stronger, and more rambunctious as his strength is returning," she says.

But now that snow cloaks the ground, we can't continue our physical therapy in the Turtle Garden. Matt heaves Fire Chief out of his stock tank and carries him upstairs. Though it's too cold for a reptile outside, at least he can wander around the living room.

At first the Chief seems thrilled to be out of the tank, thrusting his head forward and windmilling his dinosaurian legs. But once Matt sets him down, he can't walk properly. On the slick wooden floor, his front feet slip, his back legs flail, the back of his plastron drags. I try to hold up the back of his shell and let his strong front legs carry him forward. But even with my help, his mighty front claws, so effective on the soil and grass of the Turtle Garden, can't gain purchase on the slick surface inside. I try to walk him like I would a wheelbarrow in the direction he'd like to go, but he heads straight into a wall.

Alexxia calls this "white wall syndrome"—to a turtle in the wild, the sight of a large, bright, light expanse signals the presence of a pond. Next, Fire Chief walks toward some white boxes. He stretches his head out to smell them. But they offer no whiff of wetland, no intoxicating chemical messages from plants or fish or other turtles: only cardboard.

Fire Chief tries to back up, but he can't. I pick him up and reposition him. He lurches forward toward the table holding up the big tank for the red-eared sliders, where he gets his shell wedged under a folding chair. I turn him again, but now he wants to go under the hot woodstove. All the while, his back legs aren't moving at all. It's as if he's forgotten he has them.

"He's just dragging his ass," Alexxia says. "He can't move like he wants to." Fire Chief is clearly not enjoying this. We all feel his discouragement and frustration.

Finally, with our assistance, Fire Chief plods toward the sliding glass door. Here he stops. His great head and neck stretch out. His throat pumps with an elemental, piercing hunger for the world outside. He gazes out the window toward the snow-covered Turtle Garden with a palpable longing that breaks our hearts.

It's one of the few times Matt and I drive back home sad.

"Fire Chief certainly deserves an update," Natasha dictates to me in an email just days later. "On evening rounds, I got to witness a rather rare sight. Fire Chief was losing his mind clawing at his tank. When Alexxia arrived with a banana, he all but ignored it. Our big guy has gotten a taste of his freedom and, especially with his lack of exercise since you left, that freedom had a sweeter taste than the banana." She jokes, "The next thing we will be inventing will have to be the snapping turtle elliptical machine."

She's right on one count: We have got to do *something* for Fire Chief. But what?

"If this place were carpeted, he'd be zinging," Alexxia had noted earlier. Unfortunately, carpeting the upstairs is not an option. Pizza Man and Sprockets roam the house freely most days of the week, and—though there are websites that attest that tortoises can, with difficulty, be potty trained—these two are not.

So on one visit, I bring four long, braided, machine-washable rugs purchased from a discount store, hoping they'd give his feet some grip. Only they won't lie flat beneath his claws; his first few steps just ball up the rug beneath him. (We end up using them beneath the box turtles' quarters downstairs, where the residents, to our dismay, have now started pushing open the front doors to their habitats—risking a plunge to the concrete floor.)

We discuss our options. Though other recovering turtles often roam about, free, over the downstairs hospital floor, these patients are all smaller than the Chief, which make them far less likely to get wedged somewhere and dislodge filtration and plumbing equipment and flood or drain other tanks (though this has happened). And most of these other turtles either are able to hold up their own bellies above that floor's hard surface, or move so little that there is scant risk of scraped plastrons.

Could the Chief at least slide along if we made him a sort of sled to lift and cushion the back of his plastron—perhaps with a slippery piece of plastic glued to a folded towel somehow attached to his belly? Is there some way we could still allow him to use his back legs, but with a little extra support—the way a walker helps a person walk when recovering from hip replacement or a stroke?

What about a sling? Jeanne, Matt, Michaela, and I sort through various different fabrics people have donated in the bin reserved for "ratty towels"—too far gone to line turtle hospital boxes, but still too good for thrifty New England Yankees to throw away.

There are sheets, towels, and a child's blue T-shirt that says "Los Angeles." The shirt looks exactly Fire Chief's size, but we resist the urge to dress him. We settle on a towel instead. We tear it in two long strips, tie them together, and position the knot in the middle of Fire Chief's plastron, so he is resting at least some of his weight on all four legs. Matt and I each hold one strip up as the three of us walk together over the living room floor.

The design is not perfect. It demands both Matt and I pay close attention to the precise amount of tension on our end of the towel, which of course changes with every step our friend takes. But the knot won't stay in the right spot under the plastron for long and keeps slipping. We have to reposition the towel every time it does.

"Would it help to have a hole for his tail and back legs?" asks Jeanne. "What about a hole for his tail, back legs, *and* front legs?" suggests Michaela. With shears, she fashions a new sling out of a sheet. But the holes are too big. The Chief's weight rips the fabric and his back legs tangle in the shreds. We go back to the original design.

Despite its flaws, with the sling, Fire Chief can now travel rapidly, and in the direction of his own choosing. His back legs don't seem to be moving much. But at least his front legs are getting some exercise, and he's not damaging his plastron or his back feet. Most importantly, he is clearly having a good time. His head is out, his eyes are bright, and he is eagerly exploring the living room. "He's doing really well!" says Jeanne. "He's enjoying it!" Matt agrees.

The Chief guides us along as he passes the couch, turns right to pass the hot woodstove, and traverses the pink-and-blue tiles of the kitchen. He rests periodically, head out, catching scents, and then he crosses the living room again. Finally he arrives at the sliding-glass door and stops to stare outside.

Last week's snow has melted. The temperature has soared to a freakishly warm 40 degrees. Natasha says he can come out and stay for fifteen minutes before he'll get chilled. We carry him out to the deck and place him on the ground.

Here, he stands tall, looking victorious. He seems to inhale the joy of being outside again for a moment—and then, he starts walking. Now both his back legs are moving. Though he rocks from side to side with each step, he is traveling on his own. And he is clearly happy.

Alexxia considers what to do next. "The sling helped him have a good day," she says, "but we need something more."

We'd like to continue to build strength and coordination in the Chief's back legs. We don't want to let them atrophy. But his attitude, Alexxia reminds us, is as important to consider as his muscles.

"Let's think about how he is processing this," Alexxia urges. "I don't want his rehabilitation to become his frustration."

Fire Chief had come a long way over the summer. While we want to continue to challenge him, we don't want to set him up to fail.

"His back legs might never fully turn back on," Alexxia says. "Frankly—and Natasha may disagree with me here—I doubt they ever will. But I say, let's bring back his ability to move, and let him do it on his own, even if it's just with his front legs."

But how?

We'll have to design a wheelchair for a snapping turtle.

I took Fire Chief's measurements. Matt, who built his own art studio, made sketches. I consulted with friends and neighbors with building and mechanical expertise. One, Hunt Dowse—who was once a captain of the fire department in my small town, and who repairs wooden boats for a living and restores antique cars as a hobby—suggested an axle with two small wheels attached to a frame strapped to the shell, with an extension to either side so the wheels were not directly under the plastron and in the way of the back legs. Another neighbor, Tom Shevenell, an earth scientist whose work demanded his share of MacGyvering, proposed using strips of spring steel to lift the rear portion of the Chief's plastron, support the turtle's weight, and reduce the friction between the bottom part of the shell and the ground surface.

We were surprised to discover on the Internet that a number of turtle wheelchairs have already been designed. A Florida tortoise who had been run over by a front-end loader got a rig designed by Walkin' Pets—one of over a dozen companies that make wheels for handicapped dogs. (There are even charities dedicated to helping owners pay for the adaptive devices, which cost hundreds of dollars.) At a Maryland zoo, to help a rescued box turtle with a fractured plastron, veterinarians enlisted the help of a Lego-lover in Denmark to create a four-wheeled design made largely from the plastic construction toys. (A year later, that turtle, who they called Scoot Reeves, had completely healed and was released to the wild.) At Louisiana State University's vet teaching hospital, docs designed a scooter for a particularly unlucky box turtle whose two rear legs had been chewed off, on separate occasions, by different predators. They used Lego wheels and syringe parts glued directly onto the turtle's plastron with animal-safe epoxy.

Such a design wouldn't work for Fire Chief, though. Because he spends most of his time in water, his rig would need to be easily removable. Also, a snapper's plastron is a Speedo swimsuit compared with other turtles' modest maillots. Because his abbreviated belly shield affords scant coverage, it offers a much slimmer, and thus less stable, expanse beneath which to rest a platform for the wheels. And then there is the matter of Fire Chief's fourteen-inch-long tail, over which, due to his injury, he has limited control. The wheels would need to be situated so they wouldn't run over his tail when he turned or backed up.

Fortunately, Alexxia has already had experience with designing wheelchairs for snappers. June Bug's provided a prototype. From the moment she hatched here seven years ago from a batch of rescued eggs, her back legs never worked. So Alexxia created an adaptive device from a curved piece of aluminum bent to the contours of June Bug's body, two lawn mower replacement wheels, bolts as axles, and bungee cords that attach to an anchor glued to her shell to hold the apparatus in place. Today she can happily scoot across the floor and turn on a dime.

But just like the best wheelchairs for people, those for animals should be customized for their users. Silva, a five-year-old former pet snapper with light-orange eyes and a delicate face, needed a much smaller rig than June Bug's. And Natasha and Alexxia wanted to avoid gluing anything permanently to her top shell; she likes to hide under a floating rug, and a protrusion could get caught on it.

Wildlife rehabbers are known to gather an eclectic array of supplies to be fashioned into new uses. One day we came in to find that Michaela had recently picked up some novelty finger skateboards on sale at a discount store back in September, just in case they might come in handy. Clipping off the curved front end of the skateboards let them lie flat against Silva's plastron. Supergluing the two skateboards together, yoked by pieces of a wooden fork saved from a takeout meal, created a double-wide wheelchair. The whole rig was attached to Silva's body with a Velcro band saved from bundling lettuce at the grocery store, reinforced with vet wrap. Matt and I were thrilled: When we set her on the hospital floor, she was off and running.

True, the design needed further modification. Silva was so overfed as a pet that fat on her back legs spilled out from between her top and bottom shells and dragged on the floor. She needed taller wheels. But this first prototype provided proof of concept.

Thirty minutes later, Silva was still rushing about, having traversed the treatment room, visited the box turtles' habitats, and pivoted to explore beneath the tank occupied by Sergeant Pockets, who had been basking contentedly on his dock until he heard the commotion below and dove. Silva was already balancing her weight better atop the wheels, and moving faster than she had before in her life.

"We'll get a rig to work for Fire Chief," Alexxia promises. "I just need to order some more supplies and experiment with designs. I may need a few tries, and I may need a few weeks. But we'll get there."

We have no doubt Alexxia will come through. Alexxia's mechanical and design skills are varied and impressive. One of her hobbies is restoring and rebuilding collectible motorcycles, of which she has thirty-five—some of which she's souped up to travel, like a turtle on wheels, at speeds far faster than originally intended. Her excellence at appliance repair is respected widely in her profession—a recognition achieved both despite, and because of, her being a woman.

On one of winter's slower-paced visits, Alexxia and I get a moment to just chat over a cup of tea. It's just the two of us upstairs; Jeanne and Matt are downstairs helping Natasha, and Michaela is out, working at her other job. It's now that Alexxia has a chance to talk with me about subjects that seem to have little to do with turtle rescue.

Except that they do.

We talk about her business, and some of the challenges she's faced as a woman. When she used to work for Maytag, she'd often appear at a client's door only to hear, "When is the *man* showing up to fix my appliance?"

"The biggest problem was older women," she says. "You show up and she asks, 'Can I help you?' And I say, 'I'm here to fix your fridge.' And she says, 'Oh, I don't think you can do it. It's heavy!'"

But at the end of the visit, that same woman, says Alexxia, "is giving me hugs and saying hoorah for me!"

Older men, interestingly, were often welcoming—although sometimes for the wrong reasons. "Starting out, I had a pink toolbox, a ponytail hanging down my back, and I show up at this old guy's house, looking adorable, to fix the washing machine," Alexxia tells me. "He calls his friend on the phone and he doesn't think I can hear him: 'Joe! You gotta come over here! There's a model here, and she's fixing my washing machine!' And soon enough, here comes a pickup truck and out comes his buddy Joe to take a look."

At a conference of appliance repair technicians, the event's emcee asked Alexxia to stand and address the group: What was the secret, he asked, that made her such an extraordinarily successful tech?

Alexxia is a good mechanic. Her parents ran a shop repairing restaurant equipment, and she learned a lot from them. But she suggests that one reason she is able to fix appliances so well—often making successful repairs that others could not—is that she looks at her job quite differently than most men do.

"I see the situation as a person with a problem that happens to involve an appliance," she told them. "There's a relationship there between me and the client: A five-minute conversation might tell me everything I need to fix that machine."

For instance, one woman had a problem with her oven. One

after another, repairmen showed up. "The oven's broken," the lady told them. "It seems like it's overheating." The men set the oven to 350 and stuck in the thermometer. It hit 350 and stayed there. The stove, they told her, was fine. The repairmen, Alexxia said, "all thought the woman was crazy."

The woman called the shop again, and this time the owner sent Alexxia. "How long have you lived here?" Alexxia asked. "Six weeks," replied the homeowner. Alexxia invited the woman to tell her about her earlier stove. "Oh, it was this great Frigidaire," she replied. "I had it over thirty years."

"So what have you cooked that it seemed like the new oven overheated?" Alexxia queried. The woman recited a list of meals that had burned, while Alexxia listened with attention and sympathy.

Then Alexxia understood the problem. The new stove wasn't broken or malfunctioning. What was "wrong" with it was it didn't work like the old one. "The old stove had just cooled down with time," Alexxia realized. She quickly found a solution: "I reprogrammed her new stove to cook cooler," she explained. "Her new stove was fine—but it wasn't fine for *her*—and she *wasn't* crazy."

Alexxia explains her approach to her clients is similar to the way she treats the turtles. "If you see it as a being with desires and wants and pain and suffering," she says, "it's not enough to just fix the problem. You have to connect with them. Otherwise you take a tortoise and pick him up and he'll go into his shell and stay there for an hour."

Hers is, in fact, a typically feminine approach to solving problems. Women are more likely to listen longer and more attentively than men, while men are more likely to leap instantly into action. This difference is reflected in a study of male and female brains, first reported in 2001. MRIs conducted on healthy men and women at Indiana University School of Medicine showed that when men listen, mainly the left side of the brain—the analytical side, associ-ated with spatial information and math—is active. When women listen, both sides are activated, so the right brain's creative and intu-itive powers are also enlisted.

Today Alexxia operates her own company, which a sign on her vehicle proudly proclaims: A WOMAN-OWNED BUSINESS. Few of her clients notice the blue, pink, and white decals on the company car or ask her what they mean.

Matt and I had noticed, though. Alexxia likes stickers, and we found them everywhere: on her purse, on the door, in the bathroom mirror. Not all of them are about turtles. I had once remarked on one on the back of her cell phone. It has a pink shape rather like a rocket, and underneath, announces, "Some Girls Have Penises."

Really? I wondered. How can that be?

Another time I had complimented Alexxia on the pretty blue, pink, and white striped earrings she was wearing, along with a matching blue, pink, and white striped belt. "How did you man-age to find those?" I asked. "Were they a set? They go together so perfectly."

"Oh, those are the trans colors," she answered matter-of-factly.

We'd never really talked about being trans before. And I did not, in fact, expect to do so on that day. Alexxia's candor came as a surprise, and soon I found myself embarrassed to be so clueless.

I did not know the difference between transsexual and trans-gender. I had no idea what the transition in "trans" actually entailed. And, though ashamed of my ignorance, I was afraid of addressing that problem by asking questions—because I feared speaking of such intimate subjects might offend my treasured friend.

Alexxia graciously enlightened me. A transgender woman lives as a woman but was thought to be male when she was born—vice versa for a transgender man. A transsexual person, explained Alexxia, is a transgender person who has had hormone therapy and/or surgery to make their bodies look and function more like the gender with which they identify. (This is how some women have penises. Not all trans people have their entire external anatomy remodeled, though there are various surgeries to do so. And transgender may also include nonbinary people.)

I asked Alexxia outright if there were some questions I should not pose. "Never ask a trans person 'What was your name at birth?'" she told me. This is known in the trans community as "deadnaming," she explained. Using the old name feels like you're rejecting the person's true identity—or worse. Publicly calling a trans person by the old name can "out" them, exposing them to discrimination, harassment, or even deadly violence.

So, mostly I listened. Much later, over several conversations, she would fill in more details.

"When I was a little child, to me gender wasn't something I thought about at all," she told me. "It didn't really matter. But I enjoyed playing house and stuffed animals with my sister. What my brother was up to, I didn't have any interest in. When I hung out with boys, I felt like an imposter. Most of my friends were girls. I never viewed myself as a guy." And in fact, whenever Alexxia had previously spoken about her childhood, she had always spoken of "when I was a little girl."

Science bears her out. Though some people believe that gender is entirely a social construct—created by the norms, behaviors, and roles associated with each sex—brain research has steadily shown that there are also numerous, well-documented biological differences between typical male and female brains. Males and females differ markedly, for instance, in the number of neurons in different areas of the brain and how they are connected. I later read research, published in 2020, reporting that the estrogen-receptor pathways in the brains of transgender individuals are observably dissimilar from those whose gender identity matches their chromosomes. Our sex organs differentiate when we are embryos, only eleven weeks after conception. But the changes that make our brains male or female don't happen until just before birth—"And once someone has a male or female brain, they have it," said the author of that study, the obstetrician/gynecologist J. Graham Theisen, a National Institutes of Health research scholar at Augusta University, "and you are not going to change it." Alexxia was a girl trapped in a boy's body. But she didn't know what was wrong.

"*Transgender* wasn't even a word then," Alexxia remembered. "I was experiencing severe issues. I didn't feel right. And I couldn't figure it out." Neither could her parents. But they knew their child was struggling. Growing up, her grandmother used to cut her hair. "When it would grow, I liked it better. But then my grandmother would buzz me down—and I felt horrible. I was traumatized by it." Finally her parents let her cut her own hair. She grew it out until it was seven feet long.

Growing her hair out took years. Meanwhile, she explained, "I began to make other tiny changes, slowly." She shopped for her own clothes, favoring tight jeans and bright colors like purple and pink, tailoring her outfits to her own taste. She experimented with makeup. She bought nail polish. She'd paint one fingernail one week, and the next week another—until slowly, all her nails were done. "I never told anyone I was going to transition," Alexxia told me. "I just slowly changed, and became who I was."

Coming out as trans was very different for Natasha. She told me about it one day on a long drive to pick up an injured painted turtle in New Hampshire. Natasha chose a special time and place to separately announce, to each friend and family member, that she'd discovered her true identity. To her delight, everyone in her immediate circle was welcoming and understanding.

Natasha had known she didn't fit her body since third grade. That was when her Catholic elementary school started making girls and boys wear different uniforms, and use separate entrances to enter church for Mass. "My sense of self was as a girl," she said. She began to pray to God to make her "normal."

She'd given up on that prayer by the time she was sent to an all-boys Catholic boarding school. In high school, she was bullied horribly—until one day she did something completely against her nature: She kicked a heavy tablet desk over in class. She would normally never do such a thing, she said. Most boys seemed to her mindlessly destructive, "pushing and smashing things just because they could." She knew she wasn't one of them, and didn't want to be. It just seemed, in this case, she had no other choice. Her tormentors left her alone after that.

When she started college, she began searching the young Internet—and there discovered the source of her distress: "I was a girl in a boy's body!"

At age twenty-one, Natasha initiated the months-long counseling process in preparation for the hormone therapy she needed. Her first estrogen treatment was a revelation: "It flooded me with relief. Testosterone had felt like a tornado," she told me. "The estrogen calmed that terrible storm." It got better with each subsequent dose: "It felt like medicine. It felt like healing. I was becoming more and more myself."

Unlike Alexxia, Natasha also had some cosmetic surgery to feminize her face. A generous surgeon donated his services in honor of trans people he had known. The procedures reduced the size of Natasha's Adam's apple, and resculpted her nose, which had previously been large and prominent.

Now I was shocked.

"You mean, you used to have a big nose like mine?" I asked.

"I can't see your nose," she replied.

I felt a fresh flood of affection, gratitude, and admiration for my friend at that moment. I had never been pretty enough for my elegant mother. But my appearance—to me, a lifelong source of self-consciousness, now growing more troublesome with the unwelcome changes that come with age—meant literally nothing to Natasha. Of course, she had never seen the details of my face because her retinas were failing; but more importantly, she would never judge me for what I looked like. She had learned to see the souls around her—without, like so many sighted people, being blinded by the accident of their outward appearance.

Before they began to look more like the gender they knew themselves to be, both Natasha and Alexxia were threatened and bullied.

"The world, back then, wanted to stomp you out," Alexxia told me. "Even friends asked me a lot of stupid questions. Or, I'd go grocery shopping, and somebody would walk by and yell a name at me."

Dangers far worse haunt trans people. Violence is shockingly common. The National Transgender Discrimination Survey found that forty-seven percent of transgender people had been sexually assaulted during their lifetimes. One in ten had been physically attacked during the twelve months before the survey.

But facing these dangers, Natasha told me, had an important benefit: They made her a better observer—and a better advocate for turtles. Trans people are constantly scanning those around them. They learn to recognize subtle cues to warn them of an attack. (The same is true of children of alcoholics, which I knew from my youth.) Paying attention to telling details comes in handy when dealing with those who cannot speak.

Even when not threatened with physical violence, as young women, both Alexxia and Natasha often faced protracted, frustrating, sometimes humiliating hassles over things like legally altering a name or changing the sex on a driver's license. But the women agree that their difficult journey forced them to learn how to persevere. "As a younger person, I didn't want to wait," confessed Alexxia. "Before you start taking hormones, you have to go through therapy for months. You had to be medically cleared." The hormones themselves may take years to show their maximum effect.

But so it is with healing turtles. "Having to go through that journey has taught me patience," Alexxia said. "With the turtles, you can't rush these recoveries."

The struggle has enlarged her heart. "I've always had empathy," she said. "But there's no doubt that being beat down by society, mentally, physically, and emotionally, has caused me to seek out the animal that needed the *most* help.

"Trans people get the short end of the stick constantly," she continued. "So do turtles."

Natasha agrees. "Alexxia and I have an idea of what it's like to be the underdog," she says. "We work with creatures who are in some ways a personification of a cipher . . ."—a secret code, a disguised message—"an animal that doesn't have a voice. They can't tell you what's wrong. When I am working with a turtle, I am reminded very much of the struggles I made it through to be myself—struggles I can barely define. And the fact was, like a turtle in its shell, I tried to hide this turmoil of pain and insecurity. So I gravitate to these silent animals. In a world where people want everyone to serve THEM, those of us on the bottom rung need to look out for each other."

To the turtles who come to her sick or hurt or abandoned, she can say, with the sincerity that comes with experience: "Feel free to come out of your shell. You can let your guard down with me. Things will get better. I'm here, and I understand."

The holidays come and go. Instead of Christmas, Natasha and Alexxia hold a Hibernation Celebration, complete with elaborate handmade gifts for each other (one year Alexxia framed a copper sheet on which she had punched the words to a love poem she'd written for Natasha in Braille; and for Alexxia, Natasha crafted a cover from strips of wood sanded to silkiness, for a three-ring binder to hold the pages of her poetry). We feel we're coming closer and closer to a new design for Fire Chief's wheelchair. There's plenty of other work to do, of course: Matt, Jeanne, and I assist with updating the annual record-keeping, weighing, measuring, and photographing each of the 229 currently in residence. We're building and painting new, larger habitats for the box turtles, preparing for the day they can all move to a now unfinished outbuilding called Precious's Garden, in honor of a boxie they loved and lost, and make more room for actual patients in the basement hospital.

Though the flow of new turtles has slowed from summer's torrent, there is still a steady drip of newcomers: a perfectly healthy Russian tortoise arrives, a pet who people inexplicably "had to get rid of." A couple from tony Wellesley pulls up in a Mercedes bringing a turtle with an ear infection—complaining their vet wants two hundred dollars to cure it. ("You didn't *have* the two hundred dollars—or you didn't want to pay it?" Alexxia asked them pointedly, arching one eyebrow. She confiscated the turtle.)

Sprockets and Pizza Man keep us on our toes, too: One day, my shoulder bag, having fallen over sideways, begins, to my alarm, to squirm. Pizza Man is exploring inside it—which he also does to grocery bags. Sprockets is equally curious. Another time, we find him standing on the open door of the dishwasher, possibly hoping to find something to eat there. A different day, he won't come out of his carrier. He's in another snit, Natasha explains, because in his opinion, she again unfairly took him out of his soaking tub before he was ready.

But much of our discussion and effort revolves, like a wheel, around Fire Chief and how to build the best device.

Should it have one wheel? Two? Three? What kind of wheel is best? A modern roller skate, notes Natasha (who, with the help of a small wheel at the end of Mr. Stickey, avidly roller-skates with Alexxia), is made to bend back and forth, and might allow him to have a natural gait. Or, suggests Matt, what about a base plate with bushing that allows it to flex? And how to attach the device? With Velcro? With vet wrap?

On a day in early January, Matt and I find that Alexxia has purchased a matchbox-sized monster truck to pirate its large, flexing wheels and has mounted them on a ten-inch-long, five-inch-wide wooden platform padded with a piece of rubber floor mat on which Fire Chief's plastron will rest. Without attaching it to his shell, we let him try it out, hoping his weight alone will keep the device beneath him, at least for a little while.

It raises up his hind end two full inches, and to our delight, even confronted with the slippery wooden living room floor, Fire Chief immediately propels himself forward. He's mainly using his front legs—but we note that now that they are not overburdened by the full weight of his body, he is able to move his weak back legs as well. He proceeds at a nice clip, though rocking back and forth, as I remember my Aunt Lucretia, a survivor of polio, used to walk. Ten steps carry him halfway across the area in front of the front door turtle guard and into the living room. He makes it as far as the woodstove—and then the device veers drunkenly to the left. His tail flops to the side and slaps the floor as the wheels slip from under him. He takes two more steps with his front legs, but his back legs can't handle the load. He stops in his tracks.

Alexxia decides to attach the device with a belt of vet wrap. She has to turn the turtle upside down for this, which he doesn't like at all. The big guy struggles, his mighty neck craning, trying to flip right-side up. "Ba-ba, no, little monsta," Alexxia croons. "C'mon, kid, we're trying to help you!" Natasha urges. In a jiffy, Alexxia has the vet wrap on and restores him to his upright position.

Fire Chief seems dazed. *What just happened?* "Your truck came off," Matt explains matter-of-factly, as he would to a person. Now Fire Chief is not standing tall on his front legs, so the front of his plastron is on the ground while his back end is up in the air.

Alexxia narrates his thoughts: "My ass is a little high."

But Fire Chief apparently decides he can deal with this. He moves forward again—using both front and back legs. We are jubilant. "And this is just a prototype!" Natasha exclaims.

Unfortunately, in less than a minute, he falls off the wheels again.

Alexxia and Natasha add a band for the tail to stabilize him further on the platform. Now he takes a dozen speedy steps, using both front and back legs. "From a physical therapy point of view," says Natasha, "this is working very well!" Not only does this offer a chance to strengthen his weak hind legs; it's clearly great for his morale.

"When you guys didn't come in over Christmas, he was going out of his mind," she confesses. "He desperately wanted out of his tank. He desperately wanted exercise. But now he's got his mojo back!"

What, I wonder, is Fire Chief thinking now? Does he rejoice that, somehow, at this moment, he feels whole again, his mobility restored to something like it was before the accident? Does he feel his spirit freed from the psychic as well as physical weight of his injury? Or does the very memory of his disability dissolve during these few, glorious moments of fluid movement?

He falls off again, and Alexxia adjusts the vet wrap a third time. His back legs are getting tangled in the tail band; the vet wrap around his middle is falling off. Two more steps and he has left the device behind and is scrabbling helplessly at the floor with his front claws.

Rather than flip him upside down again, which he hates, and because we are running low on vet wrap, we finish off his physical therapy session by us helping him hold up his back end with our hands. Matt, Jeanne, and I take turns, bent over at a 90-degree angle, our fingers wedged beneath the sharp edges of the back of his shell, in line to get bloodied by his back toenails. It's uncomfortable—but each of us would gladly do this all day if Fire Chief wanted.

But he's getting tired. We help him walk to the sliding-glass door, where he stops to look out the window. We discuss what would improve the design: A more flexible belt? A second belt that would better hold his tail? "I may make it a little less tall," suggests Alexxia.

"The wheels of the truck are too close together, and the whole device is too long. I don't think he liked it touching his tail," offers Matt. He agreed with our neighbor Hunt, the former fire captain: The wheels should be placed farther away from the turtle's body.

And once we get the design right—what about next year? It does not look like Fire Chief will be ready for release back to his former home. For one thing, if he falls on his back, so far it seems he can't flip himself back over—a death sentence in the wild.

But he would clearly benefit from more time outside, even on days when Matt and Jeanne and I aren't here to supervise.

Natasha floats the idea of building a greenhouse to extend the season—possibly a geodesic dome. Or a second Turtle Garden could be constructed, for the sloping plot of land off the back deck. It could even be predator-proofed without a solid fence, Matt suggests. His friends at Garden State Tortoise, a private facility dedicated to the preservation of turtles and tortoises, keep outdoor pens safe with an electric fence, raven-proof fishing line across the top, and a phalanx of Havahart traps along the perimeter.

"Just put up with us, Fire Chief," Natasha tells our friend. "We have all sorts of good things in your future."

"I think this'll be the golden one!" Alexxia announces at our next visit. "This is the best one I've come up with!"

It's her fourth design: She has superglued a foam seating pad atop an aluminum cross, with no fewer than four caster wheels she bought at Tractor Supply bolted to the end of each beam.

The casters should allow Fire Chief to make a K-turn when he runs into objects like the couch. Alexxia draws the shades upstairs so he doesn't moon for the outdoors. We do have to step in when he gets under a table, but that's because he's so big he doesn't fit. ("He's like an ottoman who walks around," says Natasha.)

Matt places him atop his new chariot. He takes off, all legs engaged. His gait is the normal gait of a snapping turtle—which Natasha describes as a "rumpy-stumpy kind of walk"—only it's taking him to his destination faster than ever before.

He walks to the corner by Ralph the wood turtle's habitat. He cruises into the kitchen. He inhales the smells beneath the oven, then over by the fridge, then by the bottom of the trash can and by Sprockets's empty soak tub beneath the kitchen window. He tools over to the big potted ficus with its fluffy fern. He investigates the red pet carrier where Sprockets likes to sleep. He hugs the wooden edges of the habitat for Peppi and his companion, Apricot.

"His front end doesn't even know his back end is broken," says Alexxia. "That's so good for him."

His tail stretches out straight in back as he completes a huge oval, like a hot rod racetrack: from the sliding-glass doors, to the front of the couch, and finally back over toward the front door. He executes a perfect K-turn and begins the loop again.

"He's doing his laps!" exclaims Jeanne.

"He's really exploring!" agrees Natasha.

"It's a triumph!" shouts Matt.

Fire Chief covers the entire large front room three times. "He's got to be feeling SO good right now. He's so confident—as confident as a healthy sixty-year-old snapping turtle should be!" marvels Natasha.

"He is a wild turtle who spent most of his life as king of his wetland," she reminds us. "We did not view him as harmless! Which is the reason why Alexxia and I were surprised how quickly and unexpectedly you and Matt established quite the rapport with him. This is something we see happen with Alexxia and me. It's quite something else to witness this between the three of you. I was not directing this. Alexxia was not directing this . . ."

"Fire Chief was directing this," Matt says. "We could see it in his eyes."

And we can see the joy in his eyes now as well.

"When we pick back up with physical therapy in the spring," says Alexxia, "he will be in good shape—and not just rebounding. This exercise is good for his cardiovascular health. It's good for his muscles. It's good for his head."

"In no small way," says Natasha, "we have really rebuilt Fire Chief. From exercise outside to today—he really is a different turtle."

Spotted turtle with fiddleheads of emerging ferns.

12.

Peril and Promise

Apricot, an elongated tortoise

Though it's below zero outside, March sunlight pours through the living room windows. The chores are done: tanks cleaned, turtles fed. This afternoon, Jeanne, Matt, Alexxia, Natasha, and I take some time to laze in the sun puddles pooling on the floor upstairs, basking like turtles in the radiance of our shared happiness.

In the past few weeks, Fire Chief's strength and confidence have soared. Thanks to his new rig, his front legs have grown so strong that he tries to climb the stairs from the basement—and does so, with a boost from Natasha. He has gained almost five pounds since we last weighed him, and Matt and I note with satisfaction he is even amassing some fat on his back legs.

There's more good news as well. We are waiting for a new snapper, Peppermint Snowshoes, to lay eggs. She was so named in honor of her heroic ice rescue in January.

Alexxia was at work when Natasha and Michaela got a call reporting a snapping turtle sitting on the ice of a shallow wetland—something turtles are known to do only occasionally in winter, usually to raise their body temperatures to heal some injury or infection. But sometimes they freeze to death instead.

After an hour's drive, the pair arrived at the wetland. Squinting through binoculars, Michaela could barely see the rounded shape of a carapace; the turtle was a football field away from shore. Refusing to let the younger, though sighted, woman take the risk, Natasha walked out on the ice, leaving Mr. Stickey behind in favor of two trekking poles. From shore, a frantic Michaela shouted directions toward a turtle Natasha could not see as she advanced across waters that were only partially frozen—toward a turtle who might well already be dead.

The ice held. The turtle was alive. Miraculously, Natasha and Michaela returned to the house that afternoon bearing the new year's Number 003—a young female snapper in remarkably good shape except for some shallow gashes on her tail and right leg, and a left eye swollen shut. Weeks later, Peppermint Snowshoes had recovered enough to crawl out of her hospital tank—and into the one containing Socrates, a large adult male snapper with brain trauma. His injuries were so severe that nobody had actually ever seen this turtle move. But "apparently, Socrates was more capable than we thought," Natasha observed; recent X-rays have revealed Peppermint Snowshoes is now full of eggs.

But for the moment, Peppermint Snowshoes, Socrates, Fire Chief, and the other patients are in the basement while we are enjoying a quiet, slow visit with the tortoises upstairs. Natasha sits on the couch, holding one of Sprockets's front feet in her hand like you would hold the hand of your spouse after many happy, comfortable years together. Matt, barefoot as usual, is sitting cross-legged on the floor, holding Apricot. "She has such enchanting eyes," he murmurs. She allows her stubby, scaly legs to sprawl as he cradles her greenish-yellow plastron in his warm hands.

Alexxia is immersed in admiring Pizza Man. "You've got such a cute little tail," she tells him.

Jeanne and I are lying on our bellies on the floor, looking up at Sprockets. "The bottoms of Sprockets's feet are so interesting," Jeanne comments dreamily.

"They are very sensitive between the plates," answers Natasha.

Absorbed in the company of tortoises, our conversation floats like puffy clouds across a summer sky. We are calm and happy, comfortable and hopeful. After the ravages of national riots and insurrection, after worldwide fires and floods, after the grim milestone of 500,000 Americans lost to Covid, things are getting better. A new president speaks to our nation about mending our country and healing our planet. Vaccines against Covid were announced in December; soon we will all be protected, and return, at last, to life as it was before.

"I can't wait to have classes and Girl Scouts here again," muses Alexxia.

"I can't wait to get everyone outside in the habitat," says Natasha.

"I can't wait to get Fire Chief's legs up to speed," adds Matt.

But actually, we *can* wait. The turtles have shown us how.

An online dictionary, I find, lists several definitions of this ironic verb—an "action" word used to describe inaction: "To delay action until a particular time or until something else happens." The word is "used to indicate one is eagerly impatient to do something or for something to happen," it notes. But there are other meanings to the word *wait* that are calmer, kinder, wiser: "To pause for another to catch up." "A state of repose." *Wait* is derived from the northern French *waitier*, whose origin is related to *wake*: to become alert to; to cause something to stir to life. To wait and to wake are not opposites, but twins.

We love this moment, sprawling with tortoises in our shared sunshine—and not just because we can, finally, glimpse a longed-for future. We love now because it *is* now, and because "now" holds, at once, all of time in its fullness.

. . .

The next week, Alexxia doesn't wait for Matt to pick Apricot up out of the habitat she shares with Peppi. This time she hands Matt the tortoise—with a big red bow on her shell. Apricot is a present for Matt's fortieth birthday.

He's ecstatic. "What a year!" he says. On his last birthday, we had just returned from visiting Turtle Survival Alliance—and the next day, the world shut down. But this is not what Matt's referring to. "When Covid started," he says, "me and Erin had only four turtles. Maybe," he jokes, "Covid is not so bad after all!"

Back in September, Matt had adopted a fifth turtle from TRL, one who had been relinquished by her family. She was an adult pink-bellied side-neck turtle, a species native to Australasia. In honor of her distinctive, pinkish-orange plastron, Matt had named her Paloma, after a similarly colored cocktail he and Erin had recently enjoyed on an outing. Paloma had a bite-shaped chunk missing from the right side of the overhanging back edge of her six-inch-long carapace—where, in fact, another turtle had bitten her when she was a baby. (Though other parts of the shell can heal, edges that overhang a turtle's body do not, and the space left by the bite grows ever larger as the shell expands.)

Paloma and Apricot were just the latest acquisitions. Covid-19 had earlier caused an even bigger spike in the turtle population of the Patterson household.

When the pandemic first hit, Erin realized that, as a speech and language pathologist poking around in patients' mouths, she'd be exposed, multiple times, to a potentially deadly virus. "You could kill me!" Matt joked.

Erin started crying. Though Matt made the comment in jest, he does have asthma—and even though it's kept under control with medication, the condition put him in a high-risk group should he contract the disease.

Sometime after Matt fixed her a gin and tonic, a path to dispel

Erin's guilt became clear: If she gave Matt Covid, she promised, then he could bring home two new turtles. As month after month of the virus dragged on, her vow morphed into two new turtles just for facing the risk.

On a hot Saturday back in July, the three of us had set out to fulfill Erin's pledge. Alexxia and Natasha had told us about Faith Libardi, a rehabber in New Canaan, New York, whose nonprofit, Pet Partners of the Tri-State Berkshires, rescues turtles and dogs, and helps people in financial crisis keep their pets (and until recently, cared for a large colony of feral cats as well). Faith was desperate to find new homes for her three-toed box turtles—the same species as Matt and Erin's oldest turtle, Polly, native to the south-central United States. "They won't even need new habitats," Matt told Erin. The two new three-toed boxies could bunk with Polly and Eddie over the summer in their spacious outdoor enclosure.

How many turtles did Faith actually have? "I try not to count them," she told us upon meeting. A petite, energetic woman barely five feet tall with salt-and-pepper hair, she explained that she was working hard to find homes for the exotic turtles so she could free up space for injured native turtles coming in for rehabilitation and release. As it is, she told us breezily, she gets up at one a.m. and works till five thirty a.m. taking care of the animals before she leaves for work, running the office for a local stone quarry. A friend comes in while she's gone to give the eight dogs lunch and let them in and out to run and play whenever they wish in her large, fenced-in yard. When Faith comes home in the evening, it's all about dogs and turtles again till she drops into bed, exhausted.

Faith grew up in an animal-loving family. Her father installed windows in their barn so that, after a day of grazing, each of their dairy cows could still enjoy a nice view. The family was often late to weddings and funerals because on the way her dad would find injured or orphaned animals and stop to rescue them, no matter what species—an owl, a mouse, a turtle. There was always a box of hay in the car on which an injured animal could rest on the way back to the house for care. "As kids, we thought everyone carried boxes of hay in the trunk of their car," she told me.

But turtles have always been special to Faith. "Their souls go so deep," she said. "There's something so old and earthy about them. I never give up on them—unless they do."

She took us to the basement, where we were greeted with row after row of long, strong wooden cabinets bearing boxes full of soil, soaking tubs, hiding caves, and vegetation. Most of the turtles' enclosures are ten feet long and four feet wide, each custom-made from plate-glass-and-steel frames, and bathed in the warm glow of both full-spectrum fluorescent lights and heat lamps. Faith mentioned she spends $1,600 a year on light bulbs alone.

"Most of these turtles were confiscated," she explained. She introduced us to Megul, an eastern box turtle, who had lived for fifty years in a museum—housed, completely inappropriately, with aquatic turtles. Willow, a small, three-toed box turtle with a high-domed, deformed shell, had lived in the same museum and was suffering from eye and skin infections when Faith took him home.

Another enclosure held Mims, who "bites like a spitfire." She introduced us to Mr. T, an eastern box turtle who was blinded when he was hit by a tractor in Long Island. "They're so tough. So what does that say about what we're doing to them?" Faith said. And she showed us a Russian tortoise, like Matt and Erin's Ivan—a male who had lived with a family for years and had the run of the house, Faith told us. "He's super friendly. But one day his people didn't want him anymore. Can you imagine?"

I immediately picked him up, and he stuck out his neck to inspect me.

Many of the large enclosures hold multiple turtles. Faith steered us to one that held several candidates. She invited Erin to select her favorite first. Erin picked up a gorgeous female, over thirty years old, with connecting white speckles around her mouth. She was calm and friendly in Erin's hands. "Her name is Addie," Faith said.

"I love her," Erin pronounced instantly.

Next, Matt made his choice: He selected Willa, with orange lines on her face, a greenish-white mouth, and an exceptionally round shell. "She's very outgoing and sweet," Faith told him.

"But really," she continued, "there's not much difference between two turtles and three . . ."

Beneath the cabinet holding the enclosure, Erin flicked Matt's thigh to catch his attention. Out of Faith's sight, she flashed three fingers at him. Three turtles: OK.

Matt immediately made a counteroffer: He flashed back five fingers. Erin made a face.

"How about this pretty little lady?" Faith said, picking up a boxie with a scarred shell. "This is Cecelia. She was in a house fire. You can see it on her shell. She's a real survivor!"

"Oh gosh, we can't leave her behind," said Matt. "She was in a *fire*!" Faith handed Erin the turtle.

How could she resist?

"Well, OK," Erin said.

Faith pointed out another turtle. "And what about this one? Isn't she adorable?" This boxie has distinctive dark markings on her upper beak, looking like a smile shadowed by a little black mustache. "Her name is Louise. A lady had her for forty-five years. She's so pretty!"

"Why not?" said Matt with a huge smile.

"Oh, I guess," said Erin, looking defeated.

"And look! Can't you just take one more?" Faith urged. "There's only one three-toed left, and then I'll have this enclosure cleared out. She can't be all by herself. Can't you just take this last one?"

The fifth turtle is Pearl, named for the white dots on her lovely face.

But I couldn't stop thinking about the Russian tortoise who

had the run of a house and was then abandoned. I blurted out, "What about the Russian?" I had felt so bad for him. And he was so friendly. "I hate to leave him behind!"

"Yes," said Faith. "Can't you take the Russian?"

My husband had been adamant when I left the house: If I came back with turtles, I'd be sleeping in the street. During our more than forty years of life together, I had brought home ferrets, chickens, parrots, border collies, and a runt pig who grew to 750 pounds and lived fourteen years before he finally died of old age in his sleep. Howard had loved them all, but they had all caused havoc—our pig, for instance, learned how to open his stall to rampage repeatedly through the neighborhood, uprooting lawns and gardens. He was not infrequently led home by police (our small-town cop kept apples in the back of the cruiser for the purpose). Disasters like this usually happened while I was away in some jungle where I could not be reached for weeks. So, though it pained me, I found Howard's animal moratorium at least understandable.

Matt had a solution. "Maybe you could have him, Sy," he said, "but he could stay at our house."

I promptly named him Comrade.

At that moment, I sensed that Matt and Faith were eager to close the deal before Erin could object. Quickly we loaded five boxes of turtles into the car, filling the back seat, while Addie rode on Erin's lap. We shut the car doors.

"Oh my God," said Erin, stunned. "I don't know what happened! Matt Patterson, what have you gotten us into? I wanted TWO!"

"But this is so great!" said Matt.

"A turtle bonanza!" I added, elated. I'd taken the lid off Comrade's box and was holding him in my lap, stroking his head and neck.

"I came along to SUPERVISE!" Erin said in shock.

"Imagine what could have happened if you didn't?" Matt replied.

Back home, we settled the turtles into their new outdoor habitats. We were most interested in how the big sulcata, Eddie, would react to the newcomers. Though Eddie's best friend is Polly, a three-toed boxie, she despises Ivan, the Russian, so much that she has been known to climb a three-foot fence to try to ram her much smaller enemy.

Eddie stomped over to inspect the new crop of three-toeds. She sniffed each from head to tail, and then, satisfied with each new roommate, moved on. In their separate pen next door, Comrade and Ivan got along fine. The two males snuggled up next to each other under a tuft of grass. I left the new family to enjoy themselves together and drove home.

But at 8:04 that evening came a phone call.

"Guess what?" Matt announced with excitement. "One of the turtles is digging a nest!"

It was Cecelia, the one who had been in the house fire.

"Sy, this is a nightmare!" broke in Erin. "What if they *all* start laying eggs?"

"Yeah!" Matt exclaimed. "What if they *all* start laying eggs?"

The next day, we discovered Cecelia had dug three holes, and in one, laid two oblong eggs. Two days later, Louise started digging at six p.m. By morning, by a clump of grass near the water dish, she had laid three.

Just as hens do, turtles can lay eggs even without a male to fertilize them. To Erin's relief, Louise and Cecelia's eggs turned out to be infertile.

But thanks to Socrates's contribution, Peppermint Snowshoes's eggs are not.

Unfortunately, Peppermint's timing is off: In Massachusetts, snappers normally lay eggs in May, not March. Alexxia and Natasha decide to give Peppermint Snowshoes a shot of oxytocin to induce her to lay so her eggs can develop in their incubators.

Natasha pulls out a forty-gallon stock tank and fills it with warm water. Snappers naturally lay eggs on land, not in water, but Alexxia and Natasha have found that a "water birth" is easier and more comfortable for a mother whose labor is induced. Matt weighs Peppermint Snowshoes to determine the right dose for the injection. She's small, just under eight pounds, and though weight and size are more likely to reflect food supply than age alone, she is almost certainly younger than the usual age at first nesting, which is nineteen or twenty. "She's a pregnant teen!" Matt declares. "Please—let's call her a child bride," I say. Alexxia gives the shot of 3.6 milliliters of oxytocin in Peppermint's right thigh as I hold her. She doesn't snap or struggle. I place her gently in the warm water. She'll get a second shot in a couple of hours.

Meanwhile, we can exercise Fire Chief—and although a mound of snow remains in the Turtle Garden, it's 60 degrees out and sunny, so we can take him outside, where he doesn't need his wheelchair. When Matt hauls him from his tank and places him beneath the mulberry tree, Alexxia gasps. "He's chunked out—and he's one big son of a bitch," she says admiringly. Our big friend takes off immediately, using all four legs. This is his first time outside in months. Will his back legs be strong enough to hold his plastron off the ground? If not, will he scrape his belly shield, or cut the skin of his back legs or tail with the sharp scutes of his shell?

"C'mon, kid!" urges Natasha. Then, to us, "He's pumping those legs!"

He's moving at quite a clip—and he's not injuring his back legs or tail. He powers up the side of Mount Olympus. Matt and I stand by in case he flips. He does not. He strides confidently down the other side and heads for the fence, full of energy, keen for the feel of soil and leaves and the scent of the world outside.

"Remember last year on his first outing?" says Matt. "He was regularly having to stop and rest."

But he clearly doesn't need to now. Stomping away with chelonian gusto, Fire Chief heads over to the ice patch—turtles, Natasha reminds us, confuse white for open wetland—but quickly discovers it's cold, and turns away. Then it's back down the fence line. And now he's holding his plastron high!

Periodically, one of us runs downstairs to check on Peppermint Snowshoes. At 12:10, Natasha palpates her tail. "There's an egg in the chamber!" she announces. At 12:25, Matt finds the first egg in the water. I remove it and clean it of slime, and place it in a washcloth-lined bowl. We return to Fire Chief as we await more eggs.

He drags his legs a little as he crawls uphill through the dry leaves along the back fence. "This area has always been a big workout for him," Matt reminds us. Fire Chief turns, and heading downhill, again stands tall.

"He just needed someone who believed in his abilities," says Natasha. "And when you have so many turtles, you don't always challenge every one of them to the best of their abilities. You two believed in him."

"Yes," Matt and I say simultaneously, "we did."

Perhaps, after a summer of physical therapy, this will be the year we release him, back to the firehouse pond he had ruled for more than half a century—and where he could enjoy as much as a century more. Perhaps, with the vaccine, Covid will finally be vanquished; we can travel again, return to Turtle Survival Alliance, maybe even visit some of the exotic turtles in Southeast Asia to see them in the wild.

We check Peppermint Snowshoes for eggs again. Now there are nineteen of them—and more may be coming. We remove and clean them and place them in the bowl on the washcloth. Later they will nestle into a cushion of vermiculite in the incubator. By Sunday, there will be twenty-four, and the eggs will show dark bands indicating embryos developing inside. Because these babies won't spend the winter brumating, but growing, when they're released in early summer, they'll be the size of two-year-old wild turtles, too big for many predators to eat: twenty-four new little lives, thanks to the union of two turtles who but for TRL's efforts would both be dead.

"Eggggggs!" I cry as Matt and I get in the car to drive home. "And so early!" he replies. "More will soon be on the way," he adds.

But while eggs literally embody new life and new beginnings, even they, like every moment in time, also contain the potential for the opposite. Like all beginnings, eggs can be fraught with danger—as we would discover shortly after Matt brought Apricot home.

Matt has constructed a spacious wooden enclosure for Apricot, complete with soaking pool, hiding cave, deep soil, live plants, and full-spectrum heat lamp to mimic the conditions of the warm tropical forests throughout Asia, home of her kind. Erin has fallen in love with the new tortoise. She cooks her chicken, and serves it along with fresh strawberries, blackberries, and greens. They have renamed her; they don't think of her as a fruit that we normally encounter all dried up like a prune. They call her Lucy, which means "light."

But something is decidedly wrong. Lucy isn't eating. She doesn't venture into her soaking pond; Matt has to put her there and take her out. In fact, Lucy hardly moves at all.

Realizing that Lucy needs more sophisticated care than TRL can offer, Matt takes her to a vet who specializes in birds and reptiles, a forty-five-minute drive from their house. X-rays reveal the problem: Nearly half her eleven-inch-long body cavity is taken up with three oval eggs. One is the size of a small chicken egg; another is small and rounder, about the size of one a pullet would lay; a third is jumbo, twice as large as the smallest. An exam of Lucy's fecal matter shows she has a common parasite. She is also dehydrated. She is too stressed right now to be treated for the parasite, the vet says.

He gives her IV fluids and a shot of oxytocin in hopes she will lay her eggs. When she gets home, Lucy seems to feel better; she walks around in her pen for the first time. But she doesn't dig. She doesn't lay. Matt takes her back to the vet. She gets a second shot, and later, a third.

Finally, one night Matt notices her digging a hole in her habitat as he and Erin get ready for bed. In the middle of the night, he wakes up to find her laying. "She covered it so beautifully I never would have known it was there, otherwise," he tells me in the morning. But when he digs up every inch of the soil in her habitat, he finds only two eggs.

The third egg is stuck.

This condition can be fatal. It's called dystocia, and may strike other egg-laying animals including snakes, lizards, and birds. This is what happened to Peaches, who used to be Pizza Man's female red-footed companion, and one of Alexxia and Natasha's favorite tortoises.

Peaches had been discovered, abandoned, in a commercial greenhouse, suffering from an upper respiratory infection that the couple quickly cured. She soon became what Alexxia called "a welcome little breath of femininity," an antidote to the "dudefest" of Sprockets and Pizza Man. Peaches was delicate and dignified. She even smelled sweet, Alexxia said (except for the day she found and ate a dead mouse in the Turtle Garden). Peaches slept next to Sprockets, but snuck into Pizza Man's habitat to visit him, too—which resulted in two eggs that Peaches was unable to expel. She went in to a trusted vet for emergency surgery. But during the operation, the tube that carries urine from the kidney to the bladder was severed—an injury that no turtle can survive. Peaches was euthanized on the table. Her death hit them so hard that, though she died more than two years ago, the couple still can't face her burial. Her body still lies in the TRL freezer.

Matt and I consult with our friend, Dr. Charlie Innis, the veterinarian at New England Aquarium who also oversees care of the rescued sea turtles at the aquarium's Quincy hospital. ("He's a turtlehead" was how the aquarium's octopus keeper first described Charlie to me, leading me to believe he was either scaly or bald—of which is he neither.) He recommends that Lucy visit his colleague, Dr. Patrick Sullivan, at the MSPCA-Angell West animal care center in Waltham. An extensive exam—one that lasts so long that Matt, unable to stay with Lucy due to Covid and bored waiting in the parking lot, spends the day searching for Italian wall lizards at Fenway Victory Gardens—reveals Lucy has a number of other problems, including infections in both eyes. But the worst of them is metabolic bone disease.

This is a disorder caused by improper care—no surprise from the kind of owner who would abandon a reptile in a box in the street on a sleeting day in November. Metabolic bone disease is a mineral imbalance that weakens and deforms the shell, plastron, skeleton, muscles, nerves, and—of course, since they are largely composed of calcium—eggs. Because turtles are so stoic, the symptoms of metabolic bone disease often go undetected until it is too late.

Lucy is too sick to lay her last egg.

They may have to operate. But right now, Lucy is too sick to risk surgery.

Dr. Sullivan keeps her overnight. He gives her IV fluids, antibiotics, a calcium shot. The bill for all the diagnostics, treatments, and overnight care comes to over a thousand dollars. Due to a mix-up, the hospital sends the invoice not to Matt, as he had specifically and emphatically requested, but to Erin. She receives the unwelcome news at the worst possible moment, while at work. Matt and I are on the phone when she calls him from her cell: "Matt! The hospital sent me the bill for Lucy! Holy crap! I am losing my mind! I can't believe this! I'm going to kill you!"

A week later, Lucy is still too sick to eat. But Matt and Erin are leaving on vacation. Before Covid, they had planned an international trip; they then downsized to Hawaii, and even so, they had postponed the trip for a year. Now they have gotten their vaccines and feel safe to travel—before their tickets expire.

Matt and Erin's parents, who both live nearby, will care for their dogs, cats, and, except for one weekend when I will take over, eleven of their turtles. But Lucy requires special care. She needs to stay warm: 75 degrees Fahrenheit by day, and with her light off at night, her temp shouldn't drop much below seventy. She needs daily tub soaks, twice-daily eyedrops, and injections of antibiotics in a back thigh every seventy-two hours.

She will come to stay with me.

A turtle at our house at last!

I had assured my husband that our guest would only stay a week, could not make us sick, and would not ignite a police chase by escaping her enclosure. By the time Matt arrives with Lucy, Howard has helped me clear a space in my office beneath a window, so she can enjoy the spring sunlight, and is ready with a heavy-duty orange extension cord for her full-spectrum heat lamp. It's essential for her healing that she stay warm.

Lucy comes with an assortment of accessories: Besides her smaller, lighter travel habitat—a plastic bin the size of a file cabinet outfitted with dirt and hiding cave—Matt has brought all her meals, pre-cut and individually packaged by Erin in plastic baggies, her eye medicine, her antibiotics with extra needles and syringes, her light, and her soaking tub. Erin has also made up a checklist to make sure I give her the shots on the right days, and on which I can record the time. Matt shows me how to give her the injection.

Matt and I discuss how to make Lucy's visit more enjoyable. Recently, he tells me, he took her outside during a warm May thunderstorm, and for the first time, she stuck her head out of her shell and pumped her throat, clearly enjoying it. I plan on taking her outside with me on days the mercury crawls above 70 so she can enjoy the rays. I'll give her soaks twice daily. I'll talk and sing to her. I'll stroke her head—if she ever sticks it out again. "She really doesn't move much at all," Matt tells me. "She's not eaten anything, either."

"Don't worry about a thing," I reply. "She'll be fine!"

I'm petrified she will die.

But I am so grateful to be able to hold this beautiful, calm, sweet tortoise in my hands every day. When I slowly, gently lift her up for her tepid soak, and when I tilt her to first one side, then the other, for her eyedrops, she doesn't pull in her head. In two days, in fact, she sticks out her head as if to greet me. I hope she enjoys the warmth of my skin. I seem to be gaining her trust. But her appetite and energy do not improve. I carefully arrange her food on her little plastic saucer in what I imagine is an appetizing pattern. I wiggle bits of chicken around near her face, trying to simulate the appealing movement of a tasty worm or slug. On warm, sunny days, I take her outside to sit in the grass and enjoy the sun-dappled shade. Day after day, Lucy's fresh food goes uneaten. I toss her chicken in the garbage and dump her greens and fruits in the compost. Day after day, she hardly moves.

And then, on our third day together, it's time for her injection. What if I screw this up? If I inject it wrong, the lifesaving antibiotic won't have a chance to work. What if the shot hurts her and she never trusts me again? What if the stress of a relative stranger inserting a needle in her leg sends her over the edge?

I enlist the help of a retired MD. "This is my first turtle patient," confesses my friend and neighbor, Jack McWhorter. His specialty was rheumatology—but he feels confident he can handle this case. Due to Covid, the good doctor doesn't come inside our house. We perform the procedure on the back porch. I take the vial of antibiotic from the freezer and warm it under my arm. The doctor holds the tortoise in his sure, gentle grip. I extend her left rear leg with my left hand and with my right insert the needle between two scales, and push the syringe's plunger. I rub the injection site after the shot.

She doesn't seem to mind at all. And after a few days, she begins to move—just a little. She's still not eating. But I detect some whitish paste in her soaking tub when I remove her from her daily bath. I'm familiar with this from my experience with birds—who, like reptiles, have just one opening, the cloaca, for excreting wastes. This is uric acid, the end product of protein digestion, which suggests that Lucy's alimentary canal might be reawakening at last.

And then, disaster strikes.

My husband and I wake up one morning feeling cold. We look at the electric alarm clock. Dead. We try to flick on the light. Nothing. A violent storm—which we are seeing with greater frequency with climate disruption—has knocked out the power all over town. The temperature outside is 43 degrees. Inside, it's barely 60—and falling.

Normally this would just mean donning an extra sweater and reading by a window till the power comes back on. One year, after an ice storm, our power was out for a week. But this is different. We have an ailing tropical reptile in the house!

Luckily, my cell phone still works. I call my best friend, Liz, in the next town over. "Do you have power?" I ask frantically. I am not calling out of concern for my elderly bestie. I am calling because I remember she has an automatic generator and I need to bring a sick turtle over to her house—a house which, I happen to know, has radiant floor heating throughout.

Twenty minutes later, Liz and I are both standing in the warm drizzle of her bathroom shower with the tortoise on the heated tiles at our feet. To our delight, slowly Lucy extrudes her head and neck from her shell, basking in the water's spray just like she would in the tropical Asian rainforest of her ancestors.

Liz's phone rings; it's Howard, who has set up our far more primitive generator and is heating the house for Lucy's return. Liz and I enjoy watching Lucy crawl around the heated floor—just a few steps, but the most I've ever seen her take. She seems even more active the next day. But she's still not eating. I remove the dish at night; our border collie has discovered the chicken in there. Plus, like everyone else who lives in an old New England farmhouse with no cat, we have mice.

I consult with Alexxia and Natasha. "Try something juicy, something really sweet," Alexxia said, "maybe something red." Lucy already has strawberries. I buy a watermelon. It's one of my favorites, too.

"Mmmmmm!" I say to Lucy. To illustrate its juicy sweetness, I bite into a slice, chewing loudly and with an open mouth so she can smell my breath. I place a chunk on her plate. And that night, I remove the chicken but leave the watermelon there.

In the morning, the fruit is gone.

"I thought the mice were eating it," I email Matt and Erin in Hawaii the afternoon of June 4. "But today, I actually SAW her eating her watermelon—and her lettuce!"

Lucy eats with greater zeal each day after that: not just the watermelon and greens, but the blackberries and melon and chicken, too. She's moving around. When I garden outside, I erect our dog's old mobile puppy pen to keep her from wandering away.

Matt and Erin return June 7, and take her home June 8. It's not a moment too soon to get Matt back to TRL, where the action is fast and furious: We are up to Number 108; all twenty-four of Peppermint Snowshoes's healthy babies are, along with their mother, ready for release. Alexxia and Natasha have added two fifty-quart coolers to the battery of egg incubators, bringing the total number of "Monsta Makers" to seven, and they are holding more than 150 eggs. And the Torrington site is crawling with nesting mother snappers and paints; several wood turtle nests have already been protected; and the first Blanding's of the year just nested the day before.

Matt brings a rejuvenated Lucy to Angell for another exam. She is healthy now: her eye infection healed, her appetite back, her lethargy gone. But if the egg remains stuck inside her, it will break or decay, resulting in an infection called egg peritonitis, which is almost invariably fatal. The doctor gives her one last chance before he feels he will have to resort to surgery: another shot of oxytocin. Then, under low lights, she spends the night of quiet stillness in a warm container of moist soil.

On the last day of June—as Covid cases fall to a yearlong low, as painted turtles pile on basking logs in the Torrington River, as Fire Chief paces the Turtle Garden with new confidence and strength—in her dark room in the animal hospital, Lucy is delivered of her final egg.

Our burdens lifted, our lives open to the healing summer.

13.

Sweet Release

Releasing a baby snapper

By now, we have established a ritual with Fire Chief. The first thing we do when we arrive each week at TRL is feed the colossal snapper—by hand.

I hold the Chief's favorite treat, a one-and-a-half-inch-long nugget originally formulated to feed zoo monkeys, between my thumb and forefinger, just beneath the water's surface in front of the giant turtle's face. Fire Chief shoots out his powerful neck, opens his sharp jaws—and, with studied precision, delicately snaps the nugget in half, carefully avoiding my fingertips.

"Did you ever think you would be doing *this* with a snapping turtle?" Matt asks.

No, I most certainly did not—especially after the first time we saw him murder a banana over a year ago. "It's obvious he's consciously inhibiting his bite," I observe gratefully.

I handle the second half of the nugget more gingerly. I always let go of it and withdraw my hand a millisecond before the turtle's jaws reach the treat. But Matt, braver than I, purposely allows the

Chief's beak to actually touch his fingers as he feeds him. "I love feeling his lips," Matt says. To which, if she's in earshot, Alexxia invariably replies, "Turtles don't have lips!"

Instead, all turtles have beaks, made of dense keratin, with sharp edges for cutting meat, serrations for clipping plants, or, in some seagoing species, broad plates strong enough to crush the shells of snails and clams. A turtle's beak functions like a set of teeth. Though wild snappers do not attack people in the water, when one is eating, getting your hand near their beak is something you want to avoid. "Unless you want to join the Stub Club," jokes Natasha. "Yes, this is really a thing," she assures us—mostly among keepers and hobbyists who deal with alligators and large monitor lizards like Komodo dragons. "But if you really want to become a member," she says, "a snapping turtle can help you out."

Fire Chief's gentleness always floods us with fresh awe. He is an extraordinary individual with whom we have an extraordinary friendship. But if we ever need a reminder of the destructive potential of a turtle's jaws, we need only look to Nibbles.

His tank is next to Fire Chief's, but on a lower shelf. After Fire Chief has eaten his fill, we always toss a nugget or two Nibbles's way. Before the food even hits the surface, a dark, scaly head explodes from the water, biting with the gusto of Genghis Khan beheading some Tartars. "Nibbles punches first and asks questions later," Matt says. The only other animal I've seen feeding with such terrifying enthusiasm was a fourteen-foot saltwater crocodile who I watched seize a chicken in Bangladesh. Nibbles's sudden lunges make even Matt jump.

Five days after the summer solstice, Natasha bends down and— very carefully, as Matt and I point out the direction of Nibbles's head—lifts the feisty snapper out of his tank. Despite massive fat rolls, the turtle is supple and lively, snapping and air-swimming. But Natasha's loving grip on his shell is sure; if anything, she's proud of Nibbles's wild nature. "You don't turn your back on Nibbles!" she says. She's taking him out so that she and Alexxia can weigh him—for one last time.

Nibbles's previous weight, recorded a year ago, was 11.25 kilograms. Now he's up to 13.15—twenty-nine pounds. "And two and nine are my favorite colors!" says Natasha—to her mind, the numbers conjure red and blue. She takes this as a good portend on this life-changing day.

"Look how big you are," says Alexxia. "My God! Babydoll, you look great!"

Natasha places the lunging, whirling snapper in the largest Sterilite bin we have, and quickly, Michaela snaps the lid closed before loading him into the car. "TRL 09–001," says Natasha, "our first patient, ever. Lil' Nibbly . . ."

We have reached a milestone in the lives of both the women and the snapper. Today, the turtle they have known longer than any other will be set free.

"Is everybody ready?" Natasha asks.

For years, they were not.

It was way back when Alexxia was working for Maytag, on a customer call in summer 2009, when she first met Nibbles. She found herself in a home with a well-appointed aquarium for three red-eared sliders—and a pathetic, plastic shoebox on the floor holding an undersized yearling snapper in shallow, dirty water. The customer had found the turtle emerging from the nest at the end of the summer before. The homeowner had taken the animal as a free pet, and proceeded to provide it with a habitat and diet (nothing but dried mealworms) that were all wrong. Alexxia talked her way into taking the turtle home with her; she and Natasha immediately spent "two hundred dollars that we didn't have" on a forty-gallon tank, lights, heater, filter, and food. "Nibbles was three and a half years old before he ate well," Alexxia remembers. "But he was so curious and sweet . . ."

Just a brief drive away from TRL HQ, Alexxia pulls over to

the side of the road. Matt shoulders the heavy bin containing Nibbles for the short walk down to the wetland. A carpet of lily pads crowned with white flowers stretches before us like the aisle runner at a wedding. Thrush-song spirals, exploding fireworks of sound. Matt spots another snapper basking on a log off to our left, as if waiting for Nibbles to join.

This is a place Alexxia and Natasha know well. "So many wrongs have been righted here," says Alexxia. "This is where we've released so many hatchlings."

"And so many turtles with head trauma," adds Natasha. Some of Nibbles's offspring, conceived while in captivity, were released here, too. But Nibbles himself has stayed at TRL for twelve years.

"We always talked of releasing him," says Alexxia. "We promised him this freedom ten years ago. We last talked about releasing him a couple of years ago."

Matt sets down the bin on the banks of the pond. A catbird calls, as if in encouragement. But this is hard to do.

For years, Nibbles had served as an ambassador turtle for TRL. The snapper was so gentle, and so used to being handled, that children could safely touch him. "We'd have him loose, wandering around a library, and little girls would be scooping him up in their hands," Natasha remembers. "So many were inspired and interacted with this turtle," Alexxia recalls, "kids who had never experienced what a snapping turtle was till they met little Nibbles." Children from the city came away feeling they had met a friendly dinosaur. Kids from the country came to understand snapping turtles are not malicious killing machines, but if left unmolested, just as harmless and peaceful as any other turtle.

But then, Nibbles got bigger. And if he was going to be released to the wild, perhaps he should be given an opportunity to "discover his inner snapping turtle." They stopped regularly handling him.

Still, Nibbles has never attacked anyone. We had seen him, in the Turtle Garden, crawl peacefully into Michaela's lap. But on the rare occasions when the staff picked him up out of his tank, he would gape and snap, and he would lunge and bite impressively in the presence of food. This natural behavior serves a wild snapper well: Usually, a person lifting a turtle from its pond is a disaster that a show of strong jaws wisely forestalls. And though snappers get much of their food from plants and from scavenging, in the absence of enough carrion, they may need to be able to quickly and effectively seize prey like frogs, insects, crayfish, and fish. For years, Nibbles clearly had all the skills, instincts, and excellent health he needed to thrive in the wild.

And yet, Alexxia and Natasha did not release him.

Matt and I could understand this. Some summers are so hectic at TRL, with emergency cases flooding in, and patients literally stacked high in bins awaiting care, that it's a wonder they manage to release as many turtles as they do. Matt and I could well imagine times that were just too busy to free Nibbles—and when cold weather came, it was too late.

But there was more to their delay. Setting a beloved creature free is a moment fraught with uncertainty, especially in a world bristling with human-caused horrors: cars, poachers, pollution, fishhooks, even cruel people armed with crossbows looking for living targets . . .

It's no wonder that some wildlife rehabbers make it a point never to attend the release of the charges they care for. To them, it's just too painful and frightening. Horrible things can happen to animals released to the wild. The first time my falconry instructor, Nancy Cowan, successfully freed a rehabilitated bird of prey, a red-tailed hawk, no more than a minute had elapsed after the healed bird flew from the carrier crate when it was hit by a car and killed before her eyes. A neighbor of mine who raised a monarch butterfly from a larva had the same experience: After its new wings dried, the adult lived free less than two minutes before it was plastered to a windshield. And these are creatures who fly—not turtles who crawl along at three miles an hour, whose brains can't even process the motion of an oncoming car.

Release contains within its very sound the echo of loss. Both release and loss ask us to let go. In the past months of the pandemic, we've all done so much of that: We let go of our normal schedules, surrendered our offices and classrooms, left behind movies and restaurants, parties and performances, vacations and reunions. Tens of millions quit their jobs. We've said so many goodbyes.

In many cases, though, psychologists reassure us that learning to let go is a good thing. A Consumer Reports survey discovered that because Covid forced people to leave old routines behind, fifty-nine percent said family mattered to them more than before. One-third of grocery shoppers decided to cook from scratch at home. Fifty-eight percent vowed to take better care of their health. More students chose to go to medical school. A United Nations survey found a global surge in volunteering for food banks, helping people who are elderly or disabled. Released from their pre-Covid routines, many rediscovered the healing power of nature, found new ways of working, and focused anew on community and connection.

The UCLA psychologist Judith Orloff wrote a book called *The Ecstasy of Surrender*, about what she calls "the grace of letting go at the right moment." It's a grace often considered the secret of successful aging, a time when so many of us clutch hopelessly at vanishing physical strength, at waning memory, at diminishing muscle tone and skin tone and the hallmarks of what our culture considers beauty, and at youthful opportunities we failed to choose. To surrender—to release—can be a second chance at life. In Sanskrit, Orloff notes, the word for "surrender"–*prandidhara*—is the same word as the one for "devotion."

For a rehabber, releasing an animal back into the wild is the ultimate act of devotion. But the moment is bittersweet. We'd seen this just the week before, the day before World Turtle Day, when

Matt and I had attended the release of eleven baby snappers. They had all hatched from eggs Matt had found in the town where he grew up, a tradition Matt and his dad have been keeping since the summer that Matt was ten. "At the pond where we used to go fishing," his dad, David, told me, "we saw a guy who found a whole nest of snappers, digging them up and breaking the eggs. He said they kill the geese. I said to Matt, 'Next year, we're getting here ahead of that guy.'" Each spring thereafter, they have rescued the eggs, removing them to a flowerpot filled with sand, where they incubated and hatched.

This year, Matt had given the eggs to Susie Spikol, a naturalist and educator at our local nature center, the Harris Center for Conservation Education. The babies hatched at her home in Hancock, and she had distributed them to other naturalists at the Harris Center so they could care for the babies as they grew over their first year, observing the protocols developed by Zoo New England's Hatchling and Turtle Conservation Through Headstarting (HATCH) program. Though headstarting has its detractors—David Carroll worried it could rob the hatchlings of a year of roaming wild and mapping their territory—because of the existential crisis now facing turtles everywhere, today it's been shown that the practice does more good than harm. Headstarting protects the babies from predators when they are most vulnerable, and they are released at a larger size than many predators can handle. Zoo New England studies of endangered Blanding's turtles reveal that headstarting increases a turtle's chance of survival by as much as thirtyfold.

We had all gathered on Matt and Erin's lawn that sunny Saturday in June. Lucy was still at Angell, but the other Patterson turtles were all in evidence: Eddie was out roaming loose on the lawn, with Jimmy, the tiny Hermann's tortoise, providing a dramatic size comparison; the three-toed crew and the Russians were in their respective outdoor pens to meet their eager visitors. The families brought their snapping turtles in buckets, in cardboard boxes, in plastic bins. They had names like Fluffy, Bumpy, Hot Cocoa, Kinks, and Wavy. Several were named Snappy.

"It's a turtle family reunion," Matt said.

But it was more—for on this day, the turtles would begin their lives in the wild, and the children and their families would say goodbye to animals who had become a daily joy for the lion's share of an otherwise bleak pandemic year.

We all piled into cars and headed to the pond, where the kids and their parents fanned out around its edges. Some waded ankle deep into the mud; others stayed on dry ground. But in every case, they looked for gentle slopes and some water vegetation in which the babies could hide. Each turtle, like each child, behaved differently.

One baby seemed content to sit in the water beside the little girl who raised him. "Can I feed him some grass?" asked the girl. "No," answered her mom, "the time of feeding them is over. Their mommy and daddy live in this pond." The little girl's lower lip trembled.

A seven-year-old watched, entranced, as three babies he released burrowed instantly in the pond's soft bottom. But even though he could no longer see his charges, the boy didn't want to leave. "I wish I could check up on them," he said.

Puffy was the shier of the two snappers raised by Teddy, nine, and Penny, five, and the turtle swam off like a shot when Penny opened her palm. Snappy, however, sat beside Teddy on the muddy bank for a full minute before swimming away. And then the turtle did something that Alexxia and Natasha have occasionally observed in releases before: He circled back. About a foot from shore, with the tiny black claws of his back webbed feet barely touching the sandy bottom, Snappy stopped. He pulled his head from the water and stared directly into Teddy's face. Tears coursed down the little boy's cheeks as he said goodbye. Finally the baby turtle swam off.

"It was surprising how connected everyone was to these turtles," Susie said afterward. "I'm so excited about the kids having that chance. Their little hearts just opened up to them!"

It was wonderful to see the children let their turtles go—but, as Susie admits, "there were lots of tears that day." Even her own son, a brave ten-year-old named David, cried.

If it was hard for the children to let "their" turtles go, how much more heartbreaking must it be for Natasha and Alexxia to say goodbye to someone they have lived with and loved for twelve years? Nibbles was a family member. He was a colleague. He was their inspiration.

"We've always seen him as one of the founding members of Turtle Rescue League," said Natasha.

"Without Nibbles, we might never have founded Turtle Rescue League," agreed Alexxia. "The more we learned about snapping turtles, the more we wanted to help them . . ."

Alexxia opens the lid to the bin. "Hey, babydoll," she greets Nibbles, her voice strained. And then to Natasha, "Do you want to do it together?"

The two women each place hands on the big turtle and lift him from the box. Nibbles thrashes and snaps. "Oh, baby," Alexxia croons. "Here you go," she says, placing him on the ground between her and Natasha. "It's all about you, honeybunch . . ."

When his feet touch the ground, Nibbles suddenly stands stock-still, head out. "What's this?" Natasha says, voicing the thoughts she imagines in Nibbles's head: "Mud on my feet?"

Both women are weeping. "We've been emotionally preparing for this release for years," says Alexxia. "He's absolutely heavy and healthy . . ."

The two women turn to each other and kiss. "We're saying goodbye to a friend we'll never see again," says Natasha, sniffing. Snappers generally spend most of their lives submerged, where Natasha and Alexxia would be very unlikely to spot him even if they searched.

But for now, Nibbles remains frozen, neck out, as if caught in the moment between an old life of safety and comfort, and a new one of adventure and freedom. "Let's give him some room," chokes out Alexxia. "He's concerned about us being so close." They back up three paces.

The big snapper slowly advances toward the water. "He's definitely taking to his new home," Natasha whispers. To the turtle, she says, "Go terrorize the swamp!"

"Everyone thought he was the most beautiful turtle," remembers Alexxia. She can still hear the voices of the children seeing a snapper close up for the first time: "'Look at the bumps on his tail!' 'Look at the twisty-goos on his neck!' 'Look at his back legs—he looks like he's wearing snow pants!'"

Nibbles powers under a fallen log. He vanishes beneath a mat of mud and dark weeds. But Natasha can still hear him moving.

"I love listening to the power. You can hear that fallen log creaking. He used to dig into his little sandbox, with nothing but his nostrils sticking up . . ."

Natasha closes her eyes for a moment and submerges too: into her memories, into the sounds of the swamp, into the moment and the place that has—rightfully, and with her own help—claimed a friend she shall never again see. I ask her what she is experiencing. "I'm feeling the silence that surrounds these bird songs and frog calls," she answers.

"All conversations," adds Alexxia, "that we don't understand."

A light drizzle begins, and we turn away from the wetland. The rain closes us off from this turtle's life, like a curtain falling on an act of a play.

Fire Chief's back legs are so much stronger now. They no longer drag; both work with every step. He is even using his tail—which he would need in the wild should he find a mate, and which would be essential for righting himself if he flips over. ("When they right themselves after a flip," explains Natasha, "it's like a corkscrew: They go head, arms, and the tail finishes it off.")

Between rushing out for rescues and releases, we take the Chief from his tank so he can roam about the Turtle Garden. We leave his rig behind; his hind claws won't slip on the grass and dirt as they do on the hard floor. Usually the Chief is joined by other snappers: Snowball, who is now active, eating voraciously, her head tilt imperceptible; Cardamom, a dark-colored snapper with enormous rolls of fat from living fifteen years with a hoarder who fed him hot dogs, due for release this summer; Silverback, an ancient and aggressive snapper weighing over forty pounds, his shell and osteoderms aquamarine with moss, with a healed wound that leaves an open hole in his throat; and a yellowy, dwarfish snapper with golden eyes who hatched here in 2013 and who needs a series of beak trims to correct his underbite. He was originally named Tiger, but because he is almost as yellow as a banana, and because his adorably grumpy demeanor and distinctive stance remind us of a favorite actor, we rename him Bananny DeVito.

Of them all, Fire Chief is by far the most active. He marches up and down Mount Olympus. He climbs with ease in and out of the little pool with its waterfall. He explores the tunnels.

He patrols the periphery of the fence, stopping every so often to fill his throat, drinking in the scents that carry the news of the world. One day we watched him drop his bottom jaw and hold his mouth open for forty-five seconds. It was not a yawn, or I'd have caught it. To Matt and me, it felt like a silent song, a supplication: "I want . . . I want!"

We humans share the big turtle's yearning. We are, all of us, longing for release: From the polarizing anger gripping our country. From the greed paralyzing action to counter climate change. From the restrictions and dangers of yet another year of Covid.

"I want so much for him to command a whole pond again!" I say to Matt.

"I want him to go free, too," says Matt. Fire Chief pauses, and we stroke his outstretched head. "But I don't want him around cars. In sixty years, he had one bad day—one bad minute! Less than a minute! And look what that did to him."

I shudder. "I can't even think about it."

But we must think about it. Turtle Rescue League confronts the thoughtless cruelty of the world every day.

"Yesterday was a hell of a day," Alexxia sighs as we gather again at the Turtle Garden to exercise the Chief and his snapper crew. "I'll let Michaela tell it."

"Let's just say it was so stressful, I burned over two thousand calories on my Fitbit," Michaela says.

The hellish day began at seven thirty a.m., as she was driving to TRL from her home in Rhode Island. The car in front of her hit a rabbit. "It smashed the bunny so badly that her unborn young were blown out of her all over the road," she said. Michaela pulled Matilda, her 2006 Chevy Cobalt, over to the shoulder. One of the babies was still alive. Michaela gathered it up in a towel and rushed to TRL, while Natasha started calling rehabbers to see who could help with an orphaned preemie cottontail.

Natasha met Michaela at the door. She'd already set up fluffy towels and a warm, heated box for the baby. But while they were organizing where to meet the rehabber, their tiny patient expired.

Michaela had hardly caught her breath when they got a call from a woman in Springfield, where the town is draining a two-hundred-acre lake complex, a flood control project aimed at repairing a dam. She reported that turtles were stranded everywhere: painted turtles crawling onto the sidewalks, snappers walking into the road, others visible in the increasingly shallow pond, struggling to escape from the water that was draining so distressingly away. Michaela was dispatched to rescue as many turtles as she could.

"So I get there fifty minutes later," Michaela tells us. "There's a paint on the road, and an onslaught of cars coming. So I just run, *screaming*, 'STOP!!!'"

The first car dodged and missed the turtle. The second driver, in a truck, didn't even slow down—"For me *or* the turtle," Michaela says. He killed it.

Michaela then turned back to survey the carnage: over a dozen squashed turtles littered the side of the road. Most were paints, though a couple were snappers. At least twenty snappers were poised to cross, fleeing the draining waters.

"I patrolled one area. They were all crossing from one spot. I'm a hundred yards away and see a snapper in the road. A huge truck is coming! I was on the phone to Natasha, and didn't even hang up. I'm screaming, 'NOooooooooooooooo! STOP!' and running right toward it."

The truck driver pulled over. By the time Michaela reached him, he had already gotten out of the vehicle—and carried the turtle across the dangerous road.

But the Good Samaritan carried it in the wrong direction. "And now he chucks the turtle over the guardrail and into the bushes—back toward the water it's trying to escape from!" Michaela reports in dismay. Once the turtle disappeared into the water, she couldn't find or rescue him.

Next Michaela spotted five more snappers about to cross—and even more were coming. Meanwhile, Natasha had phoned Alexxia, and the two of them had headed out to join Michaela, their car loaded with transport boxes and buckets. Michaela saw a paint at the very edge of the road, about to cross. "Cars are not stopping. They don't give a damn. I cannot stand to see another animal hit today! If something is getting hit," she said with steely determination, "it's going to be me."

She got to the paint in time. She kept patrolling.

Natasha and Alexxia arrived quickly, she tells us. Now Alexxia continues the story:

"I see Michaela picking up a big turtle. I see one here. I see one there. I pull a Matt Patterson and take my shoes off. There's garbage everywhere. Glass. A turtle wedged next to a tire. One's next to a waterfall. Turtles are everywhere! They're panicked—the water is dropping, they're losing their minds! The water is undercutting the shore. The water has dropped twenty feet. We're all grabbing big, angry snappers. I'm barefoot on rocks in cold water, throwing turtles up to Michaela and Natasha, who are whipping them into buckets. Now I'm putting two turtles in each bucket. Otherwise, it's loose turtles in Michaela's car. Now we took fifteen turtles—and all of them except two paints and one small snapper, could be very ancient animals.

"We're all wet. We drive three or four miles away—two turtles are crawling around in the back of Michaela's trunk, loose—to another wetland, and off-load fifteen turtles there. We go back. I'm waist-deep in the water, feeling with my feet in the mud for snappers. I'd grab one, and find they were stacked two deep.

"Now we're at twenty-two turtles. Locals are seeing a bunch of girls in water that people treat as a trash dump. What are we doing there? One guy says, 'The reason they're draining it is it's so disgusting. Addicts throw their needles in there!'"

"Good thing snappers aren't aggressive," says Matt. "They don't bite people when they're in water. But needles do."

"So my feet are cut up and I'm standing in garbage water," says Alexxia. "I smell like a snapper's butthole. I see a tail sticking out from under a hubcap. Wait—hubcaps don't have tails! Michaela pulls the turtle out and hands it to me. Now there are twenty-three turtles!"

They drove their charges to different wetlands, as far as possible from roads, waters that were not part of the complex of ponds being drained. At least fifteen of the turtles, says Alexxia, had to be over fifty years old; some had probably lived for a century. "If I had just one of those old guys in my clinic and I saved him," she tells us, "I would feel so great. And we saved twenty-three turtles that day!"

"After the sixth turtle rescue," says Michaela, "I finally felt it was canceling out all the death."

The images of this rescue stayed with me long after Matt and I left that day. Of course I envisioned the desperate turtles, the careless drivers. And it was easy to picture brave, tough Alexxia wading fearlessly into the dirty water, and strong, patient Natasha gently hoisting big anxious turtles from Alexxia's hands. But to me, the scene that stands out most was slight, shy Michaela, a youngster just out of her teens—screaming "Noooooooo! STOP!" and rushing, headlong and fearless, arms outstretched to scoop up the choiceless victims of human hubbub and hurry to rescue them from danger.

This, I thought, is how to confront the cruelty of the world. This is how we gain release from the entrapment of our own despair.

A sunny summer afternoon. Another Snapperpalooza: While Snowball cruises through one of the tunnels and Bananny DeVito buries himself in leaf litter, Fire Chief heads up the steep slope of Mount Olympus toward the little pool and its waterfall. Though his stride is much stronger, he's rocking a bit from side to side.

As he nears the top, he teeters, then topples.

Fire Chief is on his back, upside down.

The big guy writhes as if in agony. He's not; this is the normal way a turtle rights himself, but it's excruciating to watch. He arches his powerful neck, trying to push himself to one side with his head. His tail sweeps to the side. But then, after about two seconds, he simply stops—perhaps because he sees his "staff" nearby and trusts we will help him. Matt and I instantly swoop in and turn him right-side up.

"Maybe we should put him upside down sometime on purpose, to help him practice and use his tail," Matt suggests. We realize Fire Chief needs to be able to flip himself over if he is ever to be safely released. A turtle on its back for too long is a goner. Some turtle and tortoise forums allege the lungs will eventually be crushed by the weight of the other organs if a turtle is allowed to remain in this position too long. That is certainly not true of some Galápagos tortoises—for in Darwin's day, they were routinely kidnapped from their islands and stored upside down in ships' holds for months, without food or water, awaiting slaughter to feed the crew. But an upside-down turtle in the sun can die of heat, if it isn't killed by predators or swarmed by biting ants first. Turtles know this, which is why when males do battle, they attempt to flip each other upside down, a definitive win over an opponent. (Many tortoises possess special projections on the bottom shell to use in jousting. The giant plowshare tortoise of Madagascar, for instance, is so named for the scute on the plastron that projects up and out from the front legs for exactly this purpose.) Other turtles are aware of the predicament of an upside-down companion, and some rush in to help. There are many videos of captive turtles and tortoises helping flipped companions back onto their feet.

"It's very stressful to them," says Alexxia. "Pizza Man gets so upset he poops all over himself when it happens." (This may have evolved as an adaptation in the wild, as covering oneself with excrement may make a turtle a less appealing meal to predators.)

The very thought of Fire Chief being stuck upside down so horrifies Matt and me that we can't bear to wait long enough to see if he can right himself. Matt's suggestion is wise. But purposely turning our friend upside down, even as part of his physical therapy, literally makes me feel sick.

"I don't think I could stand to do it," I confess.

"Neither could I," Matt admits.

"He's going to have to be able to do this in the wild," I admonish myself. "We can't set him free unless we know he can flip back over . . ."

"Well," says Alexxia, who's been watching from the deck, "he's really not ready yet. He's improved, yes. He's much better. But his legs and tail still don't work normally. He's not ready for release now—and maybe not ever. But certainly not this summer."

Matt and I don't know whether to feel defeated or relieved. Fire Chief deserves to rule a pond again. We wish he could spend every day this summer outside, not just the few hours we can take him out to walk in the Turtle Garden. We hate to think of him confined to his stock tank for yet another long, dark winter—the equivalent of holding a person for months in a tiny room. But every smashed turtle who comes in for repair reminds us that a single car, on a typical road that sees thousands of trips each day, can in a second end a life that should last half a century, a full century, or more. Fire Chief needs a complete recovery to give him his best chance at survival in the wild.

And even that might not be enough.

The World Turtle

14.

Starting Over

Blanding's turtle

Feet buried in squishy mud, pants and shirts soaked with brown water, Matt and I are feeling our way around a pond the size of a small kitchen, searching with our fingers beneath the scrubby, spongelike roots of floating water hyacinth for the slick, hard shells of hiding turtles.

It's July, and with Covid cases, hospitalizations, and deaths plummeting thanks to the vaccine, we've returned to South Carolina, to visit Cris and Clint and our other friends at the Turtle Survival Center. Though the world of humans went haywire, it was a great year for the critically endangered turtles and tortoises breeding here. The Rote Island snake-necks, whose babies Cris once had to free from their eggs by hand ("You chip a little hole in the eggshell, and try not to pierce the membrane, and look in and try to see the eye to see if the hatchling is awake") are hatching on their own. So far this year, turtles in residence have laid more than 250 eggs—"Remarkable when you consider many of these species only lay two or three or four eggs," Cris reminds us. Species like the East

African pancake tortoise, in fact, typically lay just one egg. Both our friends have been promoted, too: Cris is now director of the Center, and Clint is its assistant curator.

Within minutes of our arrival, Matt and I have immersed ourselves in the world of these endangered turtles—literally.

We've joined TSA keepers Rachael Harff and Kelly Currier, as well as interns Lauren Otterness and Lily Kirkpatrick, in one of the center's outdoor ponds, where Clint directs us to "feel around till you feel a turtle, then catch it."

We soon discover that, as well as turtles, there are many other creatures with us in the water. Less than a minute after I hop in up to my waist, inches in front of me someone's dark head pops to the surface. It's a water snake, whose expression looks as surprised as my own: "What are YOU doing here?" The bite of this species is not dangerous, but we are warned there could be hidden copperheads, also—though nobody mentions what, if anything, to do about it. Fish and crawdads and a number of large jumping spiders share the water with us, too. But there should be, as well, somewhere in this pond, twenty-six roughly one-pound, critically endangered Chinese red-necked pond turtles, whose home range is now restricted to only two spots in China and possibly one in Vietnam.

Matt is in his element: barefoot and muddy, grabbing turtles every few seconds, he sometimes comes up with a turtle in each hand. Within ten minutes, our team has caught all twenty-six, whom we then weigh, measure, check for health. After logging their updated data into the computer system, we return them to their pond—to continue growing large and strong enough to make more pond turtles to eventually repatriate to the wild.

We repeat our actions in a second waterhole, and then proceed to a pond larger and more interesting yet. This one is the size of half a basketball court. Nine turtles of two species are living in its café-au-lait-colored water, including the Malaysian pond turtle, a contender for the largest species of hard-shelled freshwater turtle in the world, with a shell growing over two feet long.

"Feel around the edge of the pond for holes," Clint instructs us, "and stick your hands in there to see if you can find a turtle."

Sticking my hands into dark holes—where a justifiably angry, possibly venomous, and likely biting critter might be hiding—is one of the things I've always been told to avoid. That might be one of the reasons I am still alive after years of visiting deserts and rainforests, and scuba diving among moray eels, stinging corals, and venomous stonefish. Matt has been similarly warned by his mom and his wife (but not by his dad—who, when Matt was a child, once held him upside down by the feet over an alligator hole so he could pick up and examine a musk turtle). Being asked to do the opposite is strangely freeing. I trust Clint and cheerfully do as he says; Matt is positively euphoric.

A particularly good spot to look for turtles, Clint suggests, is one where the water is over my head, in a deep hole. This, he notes, used to be the favored hibernation area for an alligator named Hopey.

"Why did they name him Hopey?" I ask.

"It's for 'Hope 'e doesn't bite me,'" answers Clint.

"He's gone, right?" I ask. (He is—before Turtle Survival Alliance moved in, Hopey moved to a public reptile display.)

Groping about, Matt pulls from one of the holes a dead mouse. Rachael gets hit in the face with a leaping fish. Later she pulls a leech off her neck. The holes I'm exploring in the shallower water have yielded nothing but a banana peel, so I swim over toward Matt, in the waters over my head, by a fallen tree near Hopey's old hole.

Matt suddenly lunges, splashing up to his lips in the muddy water. "Sy! Get him!" he shouts. Huh? Where? Something large and hard torpedoes into my left thigh, leaving me the next day with an impressive, four-inch purple bruise. Matt eventually gets him, though: The big turtle, we discover, now weighs thirty-five pounds—and is on track to reach one hundred.

"Can you imagine how much fun it would be to be in the pond with Fire Chief?" Matt says to me.

We look at each other and burst out simultaneously with the same sentence: "We could *swim* with him!" We could slap on our face masks and snorkels and join Fire Chief in his grace and glory—not struggling on land, not swimming two strokes and then hitting the side of a plastic stock tank, but moving weightlessly through the water. We imagine watching him spread first the fingers of one great clawed hand to propel himself ahead, and then the other, and then using his webbed back feet as thrust propellers, now strong and free. We could actually enter *his* world, rather than help him recover from the cruelty of ours. If only he had a pond again . . .

For some time over the summer, we've been privately nursing an idea. But we want it so much we don't dare speak of it yet.

Summer seems to race by. There is, as always, much to celebrate: Scratches, fully healed, is released, Peppermint Snowshoes is returned to her pond, and though Socrates has died, his offspring joined their mother the same day she is released. Spunky, Special, and Steampunk are improving markedly. Hundreds of hatchlings and dozens of healed patients are released. Ralph the wood turtle has gone to a breeding center; Jeanne's sister has adopted two TRL tortoises, a Russian she has named Laika and a baby sulcata, Maxine—who will eventually grow to weigh more than the woman does.

A record hatch at the Torrington nesting grounds—68 paints, 47 woodies, 36 Blanding's turtles, and 765 snappers—brings the total of successful hatchlings since 2009 to 4,178. Emily is head-starting eight paints, as is Jeanne (who has also adopted, from a friend of Matt's, a charismatic Russian tortoise she has named Dionysus), as well as three baby Blanding's turtles. More baby paints are going to Susie Spikol at the Harris Center to distribute to classrooms for headstarting. And to my amazement, my husband agreed to hosting four undersized inch-long painted hatchlings at our house as well.

Before Howard could recant, I rushed to a pet store half an hour's drive away and returned with over two hundred dollars' worth of equipment: a full-spectrum light and heat lamp, heater, filter, thermometer, food, floating basking platform, and a forty-gallon plastic tank.

"I thought they were going to just sit in a bucket!" my husband, remembering the snapper eggs Matt gave to Susie, exclaimed in dismay. Howard soon fell in love with the babies, though, and helped me name them after painters: Seurat, the smallest, after the pointillist; Bonnard, the largest; and two who were the same size, Monet (who loved water lilies) and Manet, a contemporary who strongly influenced Monet's work.

At first, I am afraid they will die. I fear they won't eat. I worry they'll drown. I remember with horror the story of the beloved pet adult red-ear belonging to a friend of TRL, who was stuck underwater, one claw caught in the grate over the filter, and died in the safety of his own aquarium. But no: Each morning when I turn on their heat light, I find them extremely busy, swimming, investigating the rocky bottom of the tank, poking their heads inquisitively through the interstices between the lobes of the maple leaves, and later the lettuce and kale, that I float in their tank to simulate the cover of lily pads. Soon they are eating with reptilian avidity. They embody the confidence for which their persistent, prehistoric evolution has prepared them, hungry to embrace the wonder of the world.

But to us humans, everything feels in uncomfortable flux. Just as Covid seemed in retreat, a new variant, sixty percent more transmissible than the Alpha, has, by August, become the most dominant strain in the world. Buckling under the stress of nearly two years of a pandemic whose end seems ever elusive, people are reacting with an unprecedented spike in drunk driving, road rage, physical threats and attacks aimed at healthcare workers and school board officials, and soaring rates of depression, suicide, and drug overdoses. The simple request to wear a mask provokes tantrums: Folks scream at waiters in restaurants, and on airplanes, knock out flight attendants' teeth. Struggling with the emotional long haul of the pandemic, people report feeling stuck in a joyless, psychic netherworld, somewhere between panic and depression. The word for this, writes the psychologist Adam Grant in the *New York Times,* is *languishing*—a sense of aimless stagnation and emptiness. "And it might be the dominant emotion of 2021," he asserts.

Matt and Erin are in their own peculiar limbo. Sick of gunshots all day from neighbors who consider their yard a shooting range, tired of rants from the Holocaust denier across the street, they sold their house in New Ipswich over the summer—just before their block was overwhelmed by an ugly new housing development. Temporarily lodging at the in-law cottage of some friends, they're looking for a new place in a new town amid the tightest real estate market in generations.

And now, suddenly, it's November. On a cold, sunny Friday, eight of us stand inside the Turtle Garden, around a three-foot-square by three-foot-deep hole. Natasha and Alexxia, Michaela and Andi, Mike Henry and his fiancée, Rachel, and Matt and I have gathered, again, to bury the dead.

Unlike earlier burials we've attended, this one is for only a small handful of turtles, special individuals who had been adopted and loved. Some of their bodies have been in the TRL freezer for months, waiting for the day their people were ready to say goodbye. That day is today.

Rosie, a yellowfoot tortoise, is first. She was sick for six months before a member of the League insisted the owner get help for her. She arrived at TRL "a hollow shell," Alexxia tells us. She was immediately diagnosed with pneumonia and started on IV fluids and antibiotics. But it was then too late. Rosie lasted only twenty-four hours. Her death was an avoidable tragedy; like all deaths at TRL, it stung. But hers was not nearly as hard to face as the loss of Sugarloaf, who Mike and Rachel will next lay to rest.

When Mike adopted Sugarloaf, a painted turtle, from Alexxia and Natasha four years ago, TRL gained one of its most ardent and effective supporters. Our friend Robin Hood, the big snapper with the arrow in his neck, was just one of the many dozens of turtles who owe Mike thanks for their rescue. "So Sugarloaf has saved a lot of other turtles that way," Natasha tells us.

"She was a life-changer," Mike would later tell me.

Mike doted on Sugarloaf. "During her previous eighteen years in captivity," Natasha tells us, "she had never seen proper care until Mike adopted her. Her colors bloomed. Her personality blossomed. I've never seen a prettier turtle." And indeed, as Mike unwraps her frozen body from its shroud of cloth, despite having been dead for nine months, Sugarloaf looks vitally alive—eyes open, shell shiny, front legs poised as if to stride forward and greet us. Nobody knows why she died. Nobody knows how old she was. Nobody knows if she had some previous condition before coming into Mike's care that would compromise her longevity.

Alexxia invites Mike to say a few words. But he is weeping. "I do have a lot to say," he tells us between sobs, "but I'm not going to be able to . . ."

We're all aware there are people who would not understand Mike's grief. Many would dismiss his love for Sugarloaf as a by-product of anthropomorphism: projecting human feelings onto nonhuman animals, loving them because we want them to love us.

"We have difficulty interpreting and relating to the activities of certain animals, such as reptiles," acknowledges Gordon Schuett, a Georgia State University evolutionary biologist. But he calls this "warm-blooded chauvinism," "naiveté or ignorance," "a brand of intellectual poverty."

Some, including a number of scientists, believe that reptiles don't experience social relationships. Some claim reptiles lack personalities or emotions. Research is proving these skeptics wrong. Schuett writes in the foreword to the scholarly 2021 book *The Secret Social Lives of Reptiles*, "Reptiles, alien in many ways to mammalian psychology and behavior, [share] more similarities with mammals as well as birds than we have allowed ourselves to appreciate."

Studies published in the book recount how female black rattlesnakes establish friendships; how monitor lizards in Thailand learned to purposely startle tourists (particularly women and kids) at food stalls to make them drop their snacks so they can seize them; and how mother giant Amazon river turtles and their young call to one another vocally both before and after hatching—and call out again to each other so they can all leave together on migration. Not just acknowledging but exploring the personalities of reptiles is so scientifically important that Zoo New England is conducting a study of wild box turtles to find out how shyness or boldness (measured by how long a radio-telemetry-equipped wild turtle takes to emerge from its shell after being handled) affects range size, growth rate, and longevity.

It makes evolutionary sense that something as complex as personality should not evolve in humans de novo. It would be hard to dispute the survival value of being able to recognize and meaningfully interact with the distinctive individuals who directly affect their lives. Sugarloaf was certainly one who did.

"She knew how much we loved her," Mike manages to say through his tears. Later, when more composed, he would tell me how he knew this. Sugarloaf clearly recognized Mike and Rachel, and upon their approach would dive off her basking dock, swim to the front of her spacious tank, and bob up and down. Mike recognized this as "the turtle beg," a request for food. But because Mike works mostly from home, he had the opportunity to observe her closely over thousands of hours ("Once I got her, I basically didn't watch TV anymore"), and he began to notice other behaviors.

One of them was a game she made up. "Feeding was feeding, but it was different with red peppers," he explained. "Yellow and orange didn't count." When presented with this food and this food only, Sugarloaf's behavior was markedly different. She would grab the treat and freeze for five to ten seconds. Then she'd rotate in her tank to face Mike—"And then, *boom*! She'd take off across the tank. But if I stopped watching, she'd come back toward me, like 'Hey, look—I still have it!'" Mike and Rachel called it the pepper chase game. The turtle behaved exactly like a puppy with a stolen sock, teasing the owner to chase after her.

After they had been together two years, Sugarloaf developed a new gesture. She'd drop her head low and stay completely still and look into Mike's face. "That was not about food," he said. "That meant, 'Let's talk. I wanna know what's up with you.'"

Mike also feels that Sugarloaf was aware she was dying. She slept on her basking platform the entire three years they were together—except the last night. She was very still. When Mike found blood on her dock, he called TRL. They realized Sugarloaf needed specialized care, and told him to take her to Angell Animal Medical Center in Boston. When he took her in, there was a moment she looked him in the face. "I never got a pain sense," he told me. "But I swear, when I looked at her, [her look] definitely said 'I'm not doing well, buddy. Bye.'"

Now, at the grave site, we are all crying. "All right, Loafster," Mike manages to croak out between tears. Mike places in the grave some of Sugarloaf's favorite pond plants: arrowhead leaves, lily pads, pickerel weed. He places her body upon them.

Next up is Snuggles, who is wrapped in a green-and-black cloth tied with a green bow. "I think everybody remembers Snuggly," says Alexxia. We certainly did. We had all mourned when she died in June. She was a snapper who had been born in 2011 without eyes.

It was clear that Snuggles's mother had taken one step off a curb before a driver hit her; Alexxia had knelt by the curb and harvested fourteen eggs from the corpse using a buck knife borrowed from her dad while her surprised parents, up for a visit, waited in the car. One egg was cracked; the other thirteen hatched. And of those, Snuggles was the only one who wasn't releasable. Alexxia was desperate to communicate with this blind turtle, and eventually figured out a way: She would scratch her back and then present a treat. Alexxia read a poem that she had written about Snuggles when the turtle was younger, titled "Bring It In, and We Can Euthanize It"—the suggestion from a vet she consulted. Several lines from it will stay with me always:

> *Does she not enjoy each breath she takes? Do you need to see*
> *to breathe?*
> *There is a big brown log in her tank, she crawls on it like a*
> *monkey. It is probably not brown to her,*
> *But under her turtle claws it feels good . . .*
> *So she can't see. Does she deserve to die for that?*
> *She loves her dark little life . . .*
> *She just wants to be a turtle, a snapping turtle,*
> *And no thank you, I do NOT want to bring her in to be*
> *Euthanized. She will be just fine. . . . here with me, whether*
> *or not she can see.*

"She made me think about turtles, and wonder about turtles in a way no other turtle did," Alexxia says. "I think she had vision in her dreams. It was tough to lose her."

Peaches is next—the beloved red-footed tortoise who used to sleep next to Sprockets, and who couldn't pass her eggs. She had died, during the operation to remove them, before Matt and I even began volunteering—but for Alexxia and Natasha, saying goodbye was too hard to face, until today.

"All of these little monsters became part of our family," Alexxia says. "Here is where all turtles, sick and hurt, have a home forever." And this, I realize, is why these special turtles have been buried here, actually inside the fence of the Turtle Garden. From here, it feels as if the souls of these turtles can keep watch over the others, such as Fire Chief, lending their blessing to their recovery.

Now Natasha invites us to speak the names of those we want to remember, and to ring the bell—a replica of a nautical bell, which, she explains, sailors considered the voice and the soul of the ship— that she installed over the entrance to the Turtle Garden earlier in the week. "Ring the bell," she says, "for those who are silent. Ring as long and as hard as we want."

"Crash," says Michaela, and rings the bell. She had been the first rescue of 2020, and became one of Michaela's special projects—but even her dedication could not reverse who knows how many years of neglect or mismanagement in the hands of an owner either too ignorant or too careless to give this Asian box turtle proper food, light, and habitat to keep her healthy.

"For all the ones whose names I couldn't say," Alexxia mutters softly.

"For all the ones we didn't get to in time," says Mike.

Mike adds some strips of red pepper to the grave. To feed the dead has been a human urge for thousands of years. Food is life. Tombs of the Tang dynasty were rich with grave goods; the ancient Egyptians prepared meat mummies to nourish the human ones. Similar traditions survive today: From the Día de los Muertos in Mexico, to the fifteen days of Pitru Paksha in India, to the festival of Bon in Japan, these celebrations find us leaving nourishing and delicious food offerings for our ancestors. In Baltimore, someone recently left a bottle of cognac at the grave of Edgar Allan Poe. We want to entice our beloved dead back for a visit. We want them to be happy and well fed wherever they are, because we still love them.

We take turns shoveling earth into the grave.

"This is their final nest," Natasha says when we are finished, echoing her words from the earlier burial. "We have learned so much from them." That is certainly the case with Sugarloaf. Her adoption changed Mike's life forever. "The nesting season was starting up when I took her home," he told me later. "There were Loafs lying, run over, in pain on the sides of streets. I couldn't take it. I've gone to exhaustion many times driving turtles to the rescue, multiple times in a day. It's not an option not to do it. I still call upon the Power of the Loaf when I am already tired but heading out to help one.

"In all seriousness," he continued, "Loafsty has helped more turtles than I can count. She never reproduced, but every turtle I help across a road, or bring in for treatment, and every egg I incubate is because of her. She's kept and put more turtles out in the wild than the number of eggs she ever could have laid. Even stuck in a tank her entire life, she found a way to contribute to the survival of her species and the other types of turtles here."

And so, the dead are with us still. They continue to teach and inspire us. We love them still. We need them still—perhaps now more than ever.

A cold and beautiful January Saturday: I'm looking forward to my morning Zoom workout with friends. But first thing, as always, before coffee, before breakfast for Howard and my border collie, Thurber, and me, I make the sun rise for the baby turtles, turning on their heat lamp and checking on each one.

It's one of the great joys that get me up each day. Though they are now only the size of a quarter, Seurat, Manet, Monet, and Bonnard are all growing fast and eating with delightful gusto, especially when I supplement their turtle pellets with occasional treats of cooked salmon, canned tuna, and cooked egg white. Since I am raising them to be wild, I handle them very little; but still, they clearly know my appearance means food is coming. I love when they pop their heads eagerly out of the water to stare at my face with their yellow eyes. I am getting to know them well enough that when I place a rock at a certain spot in the tank, I know exactly which turtle will enjoy resting precisely there.

But on this morning, I can only see three of them. Monet is missing.

Frantically I check beneath each leaf of floating kale. He is not trapped under the filter. He is not stuck in the grate of the water heater. But when I raise the floating dock, to my horror, there he is, limp and pale, pinned upside down in the inch-wide bowl of a suction cup meant to hold the platform to the side of the tank—which had, freakishly, come undone.

How many hours has he been like this? Did this happen shortly after we went upstairs to bed, at nine p.m. or so? It is now six forty-five a.m. In winter, some turtles can survive underwater for months on end. But none of the conditions that would trigger the miracle of brumation were present in the babies' heated tank. Monet could have been underwater without a breath for almost ten hours.

I hold the baby turtle's limp little body on my palm. I flip him upside down to see if he tries to right himself. He does not. His neck is all the way out, completely flaccid. His legs don't move. He is cold. When I place the tip of my forefinger under his neck, I feel no heartbeat. I detect no breath. His eyes are swollen and shut.

"Oh God!" I scream. "Monet has drowned!"

My husband tries to phone Matt, but he doesn't answer. Matt's phone battery is dead.

And then I remember the revival of Snowball, and also something Matt and I had witnessed at our last trip to the Turtle Survival Center in South Carolina.

We had been inside with Cris, looking at the eggs in the incubators; the keepers and interns had been checking a series of live traps, part of an annual survey of the eight wild species of turtles that occur on the property. Though the traps are specially designed never to harm a turtle, Lily found a young snapper, perhaps two years old, whose claw had gotten snagged in the wire and who had drowned. She was heartbroken. But Rachael knew what to do: Before coming to TSA, she had worked at the South Carolina Aquarium, which hosts a Sea Turtle Care Center. Sea turtles, especially if cold-stunned, can and do drown—and Rachael had witnessed resuscitation. The scale was different: Even a juvenile sea turtle is as big as a dinner plate. The little snapper was less than six inches long. But the process was the same. Rachael brought the revived snapper to us and demonstrated how she had pumped his legs to restart his heart and lungs. We set him free, completely recovered, that afternoon.

That snapper was much larger than Monet; my little paint weighs just a third of an ounce. But I have no other option. As the baby lies on his back on a towel on the kitchen table, I gently pull and push his tiny forelimbs, then his legs, in and out, as Rachael had shown us. Periodically I hold him between my thumb and forefinger, tail up, head down, hoping water will run out of his mouth. It does not. I press gently on his little orange plastron, hoping to restart his heart. I go back to pumping his limbs again. I quickly Google "Turtle CPR": Websites advise to keep the turtle right-side up. I flip him over and continue.

After twenty minutes, Monet's neck moves—once. He cranes it as if he is upside down trying to flip right-side up. But when I was a child, my father told me that he once saw the corpse of his friend sit up in its coffin, even though he was still dead; later, I learned this can be caused by a rare electrochemical reaction in the nerves. I continue CPR. Monet still appears dead. But Natasha and Alexxia's words are echoing in my head: "Never give up on a turtle."

My husband takes our border collie, who is clearly suffering from feeling my distress, outside for his customary forty-five-minute Saturday-morning walk. When he returns, Howard finds that time has run backwards: Monet has gone from dead to alive. The turtle's eyes are open, all four legs move, he holds up his head. He seems dazed, but he can walk.

I consult Clint, Alexxia, and Natasha. I float a dry translucent plastic hospital box, carpeted with a paper towel, in the warm waters of the tank, where Monet can still see the others. From a paper plate I fashion a little igloo where he can hide if he feels scared, or wants shelter from the heat lamp. After twenty-four hours, I add shallow water to the hospital box; after forty-eight, Monet is swimming with the others again. Two weeks later, he resumes eating.

Natasha and Alexxia were right: Never give up on a turtle. Because turtles never give up.

On one of the last days of February, as the world held its breath at the brink of war in Europe, Matt and I made a pilgrimage we had long considered. We went to find Fire Chief's home pond.

We thought it would be easy. We knew the town where he'd been rescued, and thought we'd just find that town's firehouse. But it turned out to be a community of thirty-three thousand people, served by no fewer than five fire stations. We looked up each address on Google Earth until we found the one that fit Alexxia and Natasha's description. There it was, as seen by satellite: Just outside the firefighters' brick and clapboard building sat the summer pond—and across the road, the waterhole where Fire Chief had spent more than sixty winters. But even though we'd seen it on our computer screens, when we got there, midmorning on the President's Day holiday, we were still surprised.

Even enlarged by a recent rain, the heart-shaped fire pond was much smaller than we expected—less than an acre. We had thought such a giant turtle would surely need a huge pond, but no: This small one was obviously providing him with enough fish, insects, carrion, and plant foods to keep him healthy.

The satellite photo had been taken in the summer. Lily pads covered much of its surface then; now they were gone. Around the pond, we could see the dried stalks of sumac, sedges, bittersweet, goldenrod, and at its edge, cattails. A yellow bobber floated at the water's surface; people fished here, and we thought of Tortzilla and the hooks Alexxia had removed from his mouth. But what truly chilled us was seeing the state highway, with its forty-miles-per-hour speed limit and its double yellow line, only ten yards from the edge of Fire Chief's pond. We remembered the grim statistics from the State University of New York study: Even near roads that are not particularly busy, turtles face a ten to twenty percent chance of dying *each year*. The results of the Ontario study, which looked specifically at snapping turtles, were even worse. So many turtles were killed crossing the highway there that the authors predicted that snappers would soon be eliminated completely from the study area.

We stepped, as he had, off a cement curb, to a narrow shoulder demarcated with a white line. We stood, where he had, at the spot where he must have stepped into the street. We could feel the rumble of cars beneath our feet. We saw where he must have crawled in agony back up the curb, where he had dragged himself beneath the rusted metal guardrail, where he had tumbled down the steep embankment, back into the cool shallows of his summer pond. We saw where Natasha and Alexxia must have put in their kayak for his rescue.

Even on a holiday, we had to be careful crossing that street to get to the destination our friend had risked his life to reach each year over so many decades. We could see the overwinter pond from the road. But to get there, because of a stockade fence that looked like it might have been built ten or twenty years ago, we walked through two yards—yards that must have been woods for the first two or three decades of Fire Chief's life. In one corner, a small dock and a rowboat.

The hibernation pond was a little bigger, trapezoidal, its shallow bottom carpeted with fallen oak leaves, its edges fringed with blueberry and thorny tangles of blackberry and wild rose. It was surely deeper than his summer pond, so it wouldn't freeze solid in the winter, and with good cover to protect him during his sluggish and vulnerable brumation from predators like otters. "This is it. He used to hang out here," Matt said reverently.

Like devotees, we also wanted to visit the place the Chief might have hatched. And this we found, too: It was less than a quarter mile away, within sight of his ponds. Upslope from a large reservoir serving a neighboring, much larger city, now behind a six-foot-tall chain-link fence, we found perfect snapper nesting habitat: sandy upland soils with plenty of light. Perhaps the same year I was born, his mother had almost certainly crawled from the waters of that reservoir, up the slope, and dug the nest to lay the eggs from which Fire Chief and his siblings had emerged.

We imagined our friend as a tiny nestling, weighing less than a coin, hatching from a perfectly round egg, leaving the soft eggshell behind and boiling up from the earth—as we had seen snappers do at the Torrington nesting grounds. We imagined him as an infant, alone on his long, brave voyage, crawling to his summer pond, unseen by human eyes. At that time, the former mill village was mostly rural; with the town's population in 1950 fewer than thirteen thousand, the fire station had not yet been built—nor had the houses across the street. What is now a state highway was then a country road carrying only light traffic.

We imagined Fire Chief growing large enough for people to notice him. By 1970, the town's population had grown to twenty-six thousand; the fire station was soon joined by a school up the street, one of thirteen elementary schools now serving the town.

The children who might have spotted Fire Chief as a youngster are now grandparents—perhaps the grandmothers and grandfathers of the firefighters who staff the station today.

Matt and I pictured it all, fast-forwarded, in our minds. "I'm seeing him swimming around, with no cracks in his shell, with fully functioning legs," said Matt. "I'm glad we came to see it."

"It's amazing he made this crossing successfully as long as he did," I said.

"It is," answered Matt. "Just listen to the sound. Cars. Constantly. And surrounded by houses."

Matt's resolve was definitive.

"He's not going to have that."

Patches of snow still persist in the woods. Matt and I, joined by Erin and Howard, stand, as we did in November, and the year before that, in October, facing a fresh-dug hole again. But this one is much bigger than any grave. This one, excavated by a backhoe, is larger than our kitchen. It's off a little-traveled dirt road, surrounded by fields and forest, and less than a mile from our house in Hancock—where Matt and Erin have bought their new home.

The pond being dug will be Fire Chief's.

Once the sedges and mosses have taken root at its edges, after the lily pads and pickerel weed collected from our local ponds have settled, and once the pond is brimming, Alexxia, Natasha, Michaela, Jeanne, and Emily will join us as Matt lifts Fire Chief from his enormous travel box and carries him, his great clawed feet outstretched, into the shallowest part of the new pond. He'll fill his throat with the scent of sodden moss and luscious mud, and then give himself to the water as if to the arms of a lover.

And on a day when the news is full of war, not long after Covid's toll has claimed one million American lives, twelve of us, like disciples, will gather to release my headstarted hatchlings at the river at the Torrington nesting grounds. Alexxia, Natasha, and Mike

Henry, Matt and Erin, Emily and Jeanne, Jeanne's teenage daughter, Abby, and her cousin, Cambria (whose family adopted the Russian tortoise, Laika, and the sulcata, Maxine), my editor, Kate, and her husband, Froy, will join Howard and me. I'll set Manet free first, allowing him to swim from my palm. Monet will be next—now the smallest of the hatchlings, but like the others, strong and healthy and alert, and too big for many predators to swallow.

Matt will let Seurat go. He will almost immediately begin to hunt for bugs. Bonnard, the largest, will be last, now big as a coaster and heavy as a stone. He will instantly burrow into soft mud, a new sensation for him. But before we leave, we will see the three other baby turtles pause from busily exploring their new world. Each will turn toward the shore, pull their heads from the water, and look into our faces.

If fate is kind, and if some of these babies are female, I could see some of them again—though I doubt I will recognize them. In fourteen years, female paints are mature enough to mate and lay eggs. If I am still alive and healthy, I will still be helping to protect this nesting ground. I might even transport some of Manet's and Monet's, Seurat's and Bonnard's newly hatched babies to the river. I may, one day, watch them, too, swim away from my open palm.

These painted turtle hatchlings could, with luck, live another forty years in the wild. Fire Chief could live another fifty. By then, Matt and Erin, Alexxia and Natasha, will be old; Emily, Howard, and I will be gone.

Or perhaps not, given the right perspective: When Einstein's great friend, the Italian engineer Michele Besso, died, the physicist wrote the grieving widow, "Though he has left this strange world a little before me, this means nothing. For us believing in physics know the distinction between past, present, and future is nothing more than a persistent, stubborn illusion." For Einstein, his friend is still there. As the astronomer Michelle Thaller explains, if we understand the nature of the universe correctly—as a landscape, with all of time laid out before us at once, as a whole—"He's just over the next hill—he's still there, we can't see him where we are now . . . but we're on the landscape with him, and he still exists just as much as he ever has."

This is difficult for most of us to imagine; I don't claim to fully understand it. But I love living on this sweet green earth right now—and I love knowing that no matter what lies ahead, I have used some of my precious life helping to make these turtles' long lives a possibility. Whether I am still around in any form or not, I like to think of Jeanne's kids, and their kids' kids, protecting the nests of Monet, Manet, Seurat, and Bonnard and their babies; and perhaps Fire Chief, and maybe even his offspring, will one day be making friends with my neighbors' grandchildren and great-grandchildren.

I remember when I was younger seeing the world as a series of ladders and staircases and mountains, always leading higher. From childhood, I understood that my job was to hurry up and reach the top. Time was *going* somewhere—and like everyone else, I wanted to keep "up," if not get "ahead."

The problem with time is, it goes so fast, something the Red Queen explains to Alice in *Alice's Adventures in Wonderland*: "Now, you see, it takes all the running you can do, to keep in the same place. If you want to get somewhere else, you must run at least twice as fast as that!" Time is the problem with which the white rabbit contends in the first pages of the book, looking at his pocket watch and exclaiming, "I'm late!"

And with age, time really does go faster. The long-distance swimmer Diana Nyad, seventy-two, on a podcast devoted to the wisdom of elders, agreed with my friend Liz Thomas: Time, she said, "actually speeds up as you get older. It speeds up exponentially, every month, every day, every hour."

Time sails by so quickly—and ultimately fatally—that it's not surprising its passage is compared with an arrow's flight. The concept of time's arrow was developed in 1927 by the physicist Arthur Eddington. Time, he tells us, goes in only one direction. As Thomas Kitching, UCLA lecturer of astrophysics, writes in the online news website The Conversation, "In the dimension of space, you can move forwards and backwards. . . . But time is different, it has a direction, you always move forward, never in reverse."

As evidence, Kitching cites the darkness of the night sky. This was shocking to me, for that darkness seems timeless. "When you look up at the universe, you're seeing events that happened in the past," he reminds us—it takes time for light to reach our eyes. If the universe had no beginning or end, he says, the night sky would be filled with light—for an infinite number of stars "in a cosmos that had always existed" would flood the night with their brightness.

"Why," he asks, "is the dimension of time irreversible?" We don't know. "This," he admits, "is one of the major unsolved problems in physics."

Physicists generally agree that as time passes, entropy—disorder—increases in our universe. But many of these same scientists think there may be parallel universes in which time runs in different directions.

And even in our own universe, and even in our own era, there is surely more than one way to understand and experience time.

In the West, we associate *past* with "behind" and *future* with "ahead." We look "forward" to seeing our friends next week; we look "back" on years past. But this is not true in all languages. The Chinese term for "the day after tomorrow," *hòutiān*, means "behind day"; Chinese speakers gesture in front for the past and behind for the future. In Hindi, "yesterday" and "today" are the same noun: *kal*. (The tense of the verb determines its meaning in a sentence.)

Einstein's time had no arrow. Einstein's time, writes the physicist Paul Davies, "is blind to the distinction between past and future. It does not flow." But Einstein did maintain an association between time and creation.

No matter where time is headed, people throughout history the world over have sought to explain how time began. And in a remarkable number of human cultures, in the beginning, there was a turtle.

In Hindu and Buddhist mythology, the tortoise Akupara carries the world on his back, upholding the earth and the sea. From Alaska's Admiralty Island to Polynesia, people say the World Turtle laid the eggs that hatched into the first humans. In North America's Haudenosaunee, Lenape, and Abenaki creation stories, the Great Spirit creates the homeland by placing earth on the back of a huge turtle; many now still refer to North America as Turtle Island—and indeed, this is the continent boasting the most turtle species in the world. In China, the World Turtle is named Ao; the creator goddess uses Ao's legs to prop up the heavens.

Without the turtle, we are told, the very sky would fall. Turtle wisdom reminds us that even at a moment in history that feels like the end times, we can find a way to regain our connection with creation—and like the World Turtle, take our turn upholding the earth. If we have used our time wisely, when we are ready, we can at last hand that joyous, terrible, honorable, essential burden to the next generation, and the next, and the next, in a great cycle of renewal.

As I enter my midsixties, I am starting to understand from the turtles that time may not be linear after all. Perhaps time is not an arrow, a deadly weapon flying toward its target. Instead of an arrow, perhaps time is an egg. Let's make it a turtle egg—with its promise that each end might lead to a new beginning.

It's after eight p.m. but still light out. The whirling melody of wood thrushes seems to generate a cool, aural breeze, even as buzzing mosquitoes stick to our sweat. Jeanne and Emily signal to Matt and me from the first slope of the Torrington nesting grounds. They point to their left and their right: "Watch out! Be careful!" their arms say. Turtles are out looking for nests.

Two woodies on the first slope are headed toward the river, their laying complete. In the fading light, Matt and I check out whether anyone is active in an area that Emily and Jeanne call Blanding's Flat. Matt spots a painted turtle, but her shell is clean, showing no soil. She's out prospecting, but not digging.

Twilight rises at 9:04 p.m. The thrushes have finished singing; insect voices pulse around us, close as breath, and the air swells with the scent of wild rose and fern. In the gloaming, Matt and I have lost everyone: We can't see Jeanne, we can't see Emily, and we can't see any turtles. We phone Jeanne's cell. She's followed a large Blanding's to the boundary of one of the neighborhood's backyards.

When we join Jeanne, the mother turtle is now only twenty feet away. But on this nearly moonless night, all we can see of her amid the tussocks and rocks is a faint shadow of a dome. Her shell is the shape of everything else around us. "If I look away for even a moment," says Jeanne, "I'll lose her."

But the night has done more than blight our vision. Darkness, it seems, has transformed this heavy, shell-clad reptile—or, perhaps, revealed another side of who she really is. We can see her only when she moves, and when she does, she glides forward in the darkness, this big turtle somehow rendered as graceful, invisible, and weightless as a spirit.

With the turtles, we are constantly thunderstruck with wonder. "How do those baby turtles manage to make it all the way back to the river?" Jeanne asks.

"In the hot sun, and through predators . . ." says Matt.

"And through a neighborhood of humans," I add.

By concentrating intensely, we can detect the mother's movement through some raspberry bushes. She's zigzagging toward Jeanne's house. Then she disappears.

"I can hear her," says Matt. "She's digging a hole! She found a good spot!"

She is two yards away from us. We strain our ears for the scrape of claws on dry leaf. The stars are blazing now, the mosquitoes gone. I would be happy to stay here all night with her as she digs.

But she has other plans.

She takes four steps. A whisper of dry grass tells us she's only inches away now. Except for the dome of her shell, she looks serpentine: her neck outstretched, her glittering dark eyes alert. She seems to flow down a gentle slope. She opens her mouth wide, as we have seen Fire Chief do, her senses querying the night air for clues—and then, as if galvanized by what she just learned, she swerves her head right, then left, and proceeds forward, walking within six inches of me. She passes me—and then slides her cool black-and-yellow plastron across the top of Matt's foot.

Concentrating intensely, moving quickly and with determination, the mother turtle flows past Matt, turns left, moves ahead, then turns back. She's heading toward some young pines a neighbor has planted at the boundary of his property.

But where has Matt gone?

He is lying on his belly under the pines. "She's here," he breathes softly. "She walked right past my face."

Now she is purposefully navigating along the fence line toward her destination: "She is headed back the way she came," says Jeanne. "She's not going to nest tonight."

Reluctantly, we leave the nesting grounds an hour before midnight. But it's impossible to feel disappointed on this magical night. It seems as if the stars are waking up above us. The ground glitters with mica, red, green, gold. We can feel the plants breathing.

The rhythmic songs of the crickets and gray tree frogs sound to me like little clocks. But instead of ticking time away—time going, going, gone—they seem to be *accumulating* time, season after season of mystery, wisdom, and wonder. With each trill and chirp and throb, these voices are keeping turtle time, renewing the covenants that keep the world alive, and offering us the gift of eternity.

Selected Bibliography

BOOKS

Austad, Steven N. *Methuselah's Zoo: What Nature Can Teach Us About Living Longer, Healthier Lives.* Cambridge, MA: MIT Press, 2022.

Baird, Julia. *Phosphorescence: Things That Sustain You When the World Goes Dark.* New York: Random House, 2021.

Behler, John L. *National Audubon Society Field Guide to North American Reptiles and Amphibians.* New York: Knopf, 2020.

Bonin, Franck, Bernard Devaux, and Alain Dupré. *Turtles of the World.* Baltimore: Johns Hopkins University Press, 2006.

Carroll, David M. *Following the Water: A Hydromancer's Notebook.* Boston: Houghton Mifflin, 2009.

———. *Self-Portrait with Turtles: A Memoir.* Boston: Houghton Mifflin, 2004.

———. *Swampwalker's Journal: A Wetlands Year.* Boston: Houghton Mifflin, 1999.

———. *Trout Reflections: A Natural History of the Trout and Its World.* New York: St. Martin's, 1993.

———. *The Year of the Turtle: A Natural History.* Charlotte, VT: Camden House, 1991.

Crosby, Alfred W. *The Measure of Reality.* Cambridge: Cambridge University Press, 1997.

Davies, Paul. *About Time.* New York: Simon and Schuster, 1995.

De Waal, Frans. *Different: Gender Through the Eyes of a Primatologist.* New York: Norton, 2022.

Doody, Sean J., Vladimir Dinets, and Gordon N. Burghardt. *The Secret Social Lives of Reptiles*. Baltimore: Johns Hopkins University Press, 2021.

Ernst, Carl H., and Roger W. Barbour, eds. *Turtles of the World*. Washington, DC: Smithsonian Institution Press, 1989.

Fraser, J. T., ed. *The Voices of Time: A Cooperative Survey of Man's Views of Time as Expressed by the Sciences and by the Humanities*. Amherst: University of Massachusetts Press, 1981.

Grant, Jaime M., Lisa Mottet, Justin Tanis, Jack Harrison, Jody L. Herman, and Jessica Keisley. *Injustice at Every Turn: A Report of the National Transgender Discrimination Survey*. Washington, DC: National Gay and Lesbian Task Force and National Center for Transgender Equality, 2011.

Haupt, Lyanda Lynn. *Rooted: Life at the Crossroads of Science, Nature, and Spirit*. New York: Little, Brown Spark, 2021.

Higgins, Jackie. *Sentient: How Animals Illustrate the Wonder of Our Human Senses*. New York: Atria, 2021.

Hoffman, Eva. *Time*. New York: Picador Press, 2009.

Laufer, Peter. *Dreaming in Turtle: A Journey Through the Passion, Profit, and Peril of Our Most Coveted Prehistoric Creatures*. New York: St. Martin's, 2018.

Mansfield, Howard. *Turn and Jump: How Time and Place Fell Apart*. Peterborough, NH: Bauhan Publishing, 2010.

Money, Nicholas P. *Nature Fast and Nature Slow: How Life Works, from Fractions of Seconds to Billions of Years*. London: Reaktion Books, 2021.

Morgan, Ann Haven. *Field Book of Ponds and Streams*. New York: Putnam, 1930.

O'Connell, Caitlin. *Wild Rituals: Ten Lessons Animals Can Teach Us About Connection, Community, and Ourselves*. New York: Chronicle Books, 2021.

Rou, Yun. *Turtle Planet: Compassion, Conservation, and the Fate of the Natural World*. Coral Gables, FL: Mango Publishing, 2020.

Rudloe, Jack. *Time of the Turtle.* New York: Knopf, 1979.

Schrefer, Eliot. *Queer Ducks (And Other Animals): The Natural World of Animal Sexuality.* New York: HarperCollins, 2022.

Steyermark, Anthony C., Michael S. Finkler, and Ronald J. Brooks, eds. *Biology of the Snapping Turtle.* Baltimore: Johns Hopkins University Press, 2008.

Thomas, Elizabeth Marshall. *Growing Old: Notes on Aging with Something Like Grace.* New York: HarperCollins, 2020.

———. *The Old Way: A Story of the First People.* New York: Farrar, Straus and Giroux, 2006.

———. *The Harmless People.* New York: Knopf, 1959.

Whitrow, G. J. *Time in History.* Oxford, UK: Oxford University Press, 1988.

PERIODICALS

Aresco, Matthew J. "Highway Mortality of Turtles and Other Herpatofauna at Lake Jackson, Florida, USA." *UCOET Proceedings,* 2003, 433–34.

Gibbs, James P., and W. Gregory Shriver. "Estimating the Effect of Road Mortality on Turtle Populations." *Conservation Biology* 16, no. 6 (2002): 1647–52.

Healy, Kevin. "Metabolic Rate and Body Size Are Linked with Perception of Temporal Information." *Animal Behavior* 86, no. 4 (2013): 685–96.

Hoagland, Edward. "On Aging." *American Scholar,* March 1, 2022, 106.

Johnson, Albert, James Clinton, and Rollin Stevens. "Turtle Heart Beats Five Days After Death." *American Biology Teacher* 19, no. 6 (1957): 176–77.

LaCasse, Tony. "Flying Sea Turtles and Other Means of Rescue." *Natural History,* February 2019, 35–41.

Lapham, Lewis H. "Captain Clock." *Lapham's Quarterly* 7, no. 4 (2014), 13–21.

Lohmann, K., et al. "Geomagnetic Map Used in Sea-Turtle Navigation." *Nature* 428, no. 6986 (2004): 909–10.

Lovich, Jeffrey E., Joshua R. Ennen, Mickey Agha, and J. Whitfield Gibbons. "Where Have All the Turtles Gone and Why Does It Matter?" *Bioscience* 68, no. 10 (2016): 771–79.

Piczak, Morgan L., Chantel E. Markle, and Patricia Chow-Fraser. "Decades of Road Mortality Cause Severe Decline in a Common Snapping Turtle (*Chelydra seprentina*) Population from an Urbanized Wetland." *Chelonian Conservation and Biology* 18, no. 2 (2019): 231–40.

Stanford, Craig, John B. Iverson, Anders G. J. Rhodin, et al. "Turtles and Tortoises Are in Trouble." *Current Biology* 30 (2020): 721–35.

ONLINE

Angier, Natalie. "All but Ageless, Turtles Face Their Biggest Threat: Humans." *New York Times*, December 12, 2006, https://www.nytimes.com/2006/12/12/science/12turt.html.

Collins, Peter, and Juan Carlos López. "Listen Without Prejudice." *Nature Reviews Neuroscience* 2, no. 1 (2001): 6, https://doi.org/10.1038/35049024.

Fields, Helen. "ScienceShot: Hibernating Turtles Aren't Dead to the World." *Science*, October 8, 2013, https://www.science.org/content/article/scienceshot-hibernating-turtles-arent-dead-world.

Gartsbeyn, Mark. "720 Stranded Sea Turtles Were Rescued on Cape Cod This Season, Setting New Record." *Boston.com*, December 17, 2020, https://www.boston.com/news/animals/2020/12/17/sea-turtles-rescue-cape-cod-2020.

Giaimo, Cara. "The Celebrity Tortoise Breakup That Rocked the World." *Atlas Obscura*, February 13, 2019, https://www.atlas obscura.com/articles/tortoise-breakup-bibi-and-poldi.

Goldfarb, Ben. "Lockdowns Could Be the 'Biggest Conservation Action' in a Century." *Atlantic*, July 6, 2020, https://www.theatlantic.com/science/archive/2020/07/pandemic-roadkill/613852.

Grant, Adam. "There's a Name for the Blah You're Feeling: It's Called Languishing." *New York Times*, April 12, 2021, https://www.nytimes.com/2021/04/19well/mind/covid-mental-health-languishing.html.

Green, Jared M. "Effectiveness of Head-Starting as a Management Tool for Establishing a Viable Population of Blanding's Turtles." Master's thesis, University of Georgia, 2015, http://tuber ville.srel.uga.edu/docs/theses/green_jared_m_201512_ms.pdf.

Grundhauser, Eric. "Why Is the World Always on the Back of a Turtle?" *Atlas Obscura*, October 20, 2017, https://atlasobscura .com/articles/world-turtle-cosmic-discworld.

"Hatchling and Turtle Conservation Through Headstarting (HATCH)." *Zoo New England*, accessed March 30, 2022, https://www.zoonewengland.org/protect/here-in-new-england/turtle-conservation/hatch.

Kitching, Thomas. "What Is Time—and Why Does It Move Forward?" *The Conversation*, February 22, 2016, https://the conversation.com/what-is-time-and-why-does-it-move-forward-55065.

MacDonald, Bridget. "Loving Turtles to Death." *U.S. Fish & Wildlife Service* (blog), May 22, 2020, https://fws.gov/story/2021-06/loving-turtles-death.

Main, Douglas. "Turtles 'Talk' to Each Other, Parents Call out to Offspring." *Newsweek*, August 19, 2014, https://news week.com/turtles-talk-to-each-other-parents-call-out-to-offspring-265613.

Maron, Dina Fine. "Turtles Are Being Snatched from U.S. Waters and Illegally Shipped to Asia." *National Geographic*, October 28, 2019, https://www.nationalgeographic.com/animals/article/american-turtles-poached-to-become-asian-pets.

Massachusetts Eye and Ear Infirmary. "Brain 'Rewires' Itself to Enhance Other Senses in Blind People." *ScienceDaily*, March 22, 2017, www.sciencedaily.com/releases/2017/03/170322143236.htm.

Nash, Darren. "The Terrifying Sex Organs of Male Turtles." *Gizmodo*, June 20, 2012, https://gizmodo.com/the-terrifying-sex-organs-of-male-turtles-591970.

Ondrack, Stephanie. "The Turtle Trance." *The Small Steph* (blog), April 13, 2019, https://thesmallsteph.com/the-turtle-trance.

Rahman, Muntaseer. "How to Tell If Your Turtle Is Dead?" *The Turtle Hub*, https://theturtlehub.com/how-to-tell-if-your-turtle-is-dead.

"Star Tortoise Makes Meteoric Comeback." *WCSNewsroom*, October 11, 2017, https://newsroom.wcs.org/News-Releases/articleType/ArticleView/articleId/10600/Star-Tortoise-Makes-Meteoric-Comeback.aspx.

Waldstein, David. "Mother Sea Turtles Might Be Sneakier Than They Look." *New York Times*, May 19, 2020, https://www.nytimes.com/2020/05/19/science/sea-turtles-decoy-nests.html.

Wong, Brittany. "Turtle Divorce: Giant Turtles Divorce After 115 Years Together." *Huffpost*, June 8, 2012, updated November 22, 2012, https://www.huffpost.com/entry/turtle-divorce_n_1581463.

Help the Turtles

These organizations, mentioned in the text, can give you more information about turtles, and can use your help, too. Contact them here:

Turtle Rescue League
https://turtlerescueleague.org

Turtle Survival Alliance
https://turtlesurvival.org

Mass Audubon Wellfleet Bay Wildlife Sanctuary
https://www.massaudubon.org/get-outdoors/wildlife-sanctuaries/
wellfleet-bay/about/our-conservation-work

New England Aquarium's Sea Turtle Hospital
https://www.neaq.org/about-us/mission-vision/saving-sea-turtles/

Zoo New England's Hatchling and Turtle Conservation
Through Headstarting
https://www.zoonewengland.org/protect/here-in-new-england/
turtle-conservation/hatch

Pet Partners of the Tri State Berkshires (Faith Libardi's turtle
rehab, which also helps people in financial crisis keep their pets)
https://www.petpartnersberkshires.org

Call (518) 781-0362 if you have an injured turtle.

To find a turtle rehabilitator near you:
https://www.humanesociety.org/resources/how-find-wildlife-
rehabilitator

Acknowledgments

I owe a huge debt of gratitude to all the people and turtles mentioned by name in these pages, especially Matt and Erin Patterson; Alexxia Bell, Natasha Nowick, Michaela Conder, and Mike Henry at Turtle Rescue League; the Turtle Ladies (and men) of Torrington, who must go unnamed to protect the nesting grounds; and Cris Hagen and Clint Doak at Turtle Survival Alliance. Without them, this book would not exist—but far worse, neither would thousands of turtles living now, as well as their descendants as yet unborn.

But a number of others, some of them not mentioned in these pages, deserve my heartfelt thanks. Dr. Bryan Windmiller, director of field conservation at Zoo New England, has been a guiding light for turtle conservation programs for New England and beyond. His expertise was a crucial resource for this book. So were his colleagues, field biologist Julie Lisk, research associate Cara McElroy, and turtle field technician Ryan Roseen, with whom Matt and I shared a glorious day tracking box turtles as part of an ongoing study of how turtle personalities affect longevity, home range, and movements.

Matt and I are extremely grateful to the veterinarians who cared for the turtles we love. Thank you, Dr. Robert DeSena of Marlborough Veterinary Hospital and Dr. Patrick Sullivan of MSPCA-Angell West. Dr. Charlie Innis of New England Aquarium and Dr. Mark Pokras, director emeritus of Tufts Wildlife Clinic, both played much larger roles in the creation of this book than their short mentions would suggest.

I thank my friends and turtle-lovers Selinda Chiquoine, E. Fine, Joel Glick, Elizabeth Marshall Thomas, J. Urda, Betsy Small, and Gretchen Vogel for their friendship and encouragement during this project, and for reading and commenting on the manuscript.

I could not have hoped for a better editor than my longtime friend and collaborator, Kate O'Sullivan. Her sensitive and revealing comments and suggestions made this a much better book; her kindness and curiosity have made me a much better person. The same goes for my now-retired literary agent and treasured friend, Sarah Jane Freymann, with whose help Matt and I started this project, and our excellent and beloved current agents, Molly Friedrich and Heather Carr, who saw this book to its completion. I'm grateful to the careful copyeditor, Allison Kerr Miller, for her light touch and helpful suggestions.

In addition, I thank conservationist Jennifer Petit for her advice and good company, and librarian Molly Benevides for her generous help formatting the selected bibliography. To Amanda Bucchiere: Thank you for the donuts. They fueled hundreds of our trips through turtle time on the long drive from New Hampshire to Southbridge.

I feel I should also thank someone I never met: Alexxia and Natasha's chief mentor, Kathy Michell. At Fire Chief's release, Matt and I learned with great sorrow that after beating the odds for many years, she had finally lost her battle with cancer. But her good works are still walking in the world.

Finally, though my husband, Howard Mansfield, appears only periodically in these pages, he deserves credit for far more than putting up with my passion for animals. Howard is the best writer and most insightful thinker I have ever known. He read this manuscript with an eagle's keen eye. He has been my greatest human inspiration throughout my writing career. For this and so much more, I will love him to the end of time.

Index